Foundations for Interdisciplinarity in the Life Sciences: Concise Monographs

Series Editor

Alan C. Love, 831 Heller Hall, University of Minnesota - Twin Cities, Minneapolis, USA

Editorial Board Members

Colin Allen, Department of History and Philosophy of Science, University of Pittsburgh, Pittsburgh, USA

Jamie Davies, Deanery of Biomedical Sciences and Synthsys Centre for Synthetic Biology, University of Edinburgh, Edinburgh, UK

Renee Duckworth, Department of Ecology & Evolutionary Biology, University of Arizona, Tucson, USA

Walter Fontana, Department of Systems Biology, Harvard Medical School, Boston, USA

Eva Jablonka, The Cohn Institute for the History and Philosophy of Science and Ideas, Tel-Aviv University, Tel-Aviv, Israel

Elisabeth Lloyd, Department of History and Philosophy of Science and Medicine, Indiana University, Bloomington, USA

Dan McShea, Department of Biology, Duke University, Durham, USA

Gunter P. Wagner, Department of Ecology & Evolutionary Biology, Yale university, New Haven, USA

The Foundations for Interdisciplinarity in the Life Sciences book series introduces and extends scholarship addressing a growing recognition that biological phenomena which suggest agency, directionality, and goal-directedness demand new or updated conceptual frameworks that can translate into rigorous theoretical models and discriminating empirical tests. The book series provides readers with conceptual, theoretical, and experimental resources for prosecuting new kinds of inquiry across disciplinary boundary lines.

Armin W. Schulz

Presentist Social Functionalism: Bringing Contemporary Evolutionary Biology to the Social Sciences

Springer

Armin W. Schulz
Department of Philosophy
University of Kansas
Lawrence, KS, USA

ISSN 3005-1231 ISSN 3005-124X (electronic)
Foundations for Interdisciplinarity in the Life Sciences: Concise Monographs
ISBN 978-3-031-94832-9 ISBN 978-3-031-94833-6 (eBook)
https://doi.org/10.1007/978-3-031-94833-6

© The Editor(s) (if applicable) and The Author(s) 2025. This book is an open access publication.

Open Access This book is licensed under the terms of the Creative Commons Attribution 4.0 International License (http://creativecommons.org/licenses/by/4.0/), which permits use, sharing, adaptation, distribution and reproduction in any medium or format, as long as you give appropriate credit to the original author(s) and the source, provide a link to the Creative Commons license and indicate if changes were made.
The images or other third party material in this book are included in the book's Creative Commons license, unless indicated otherwise in a credit line to the material. If material is not included in the book's Creative Commons license and your intended use is not permitted by statutory regulation or exceeds the permitted use, you will need to obtain permission directly from the copyright holder.
The use of general descriptive names, registered names, trademarks, service marks, etc. in this publication does not imply, even in the absence of a specific statement, that such names are exempt from the relevant protective laws and regulations and therefore free for general use.
The publisher, the authors and the editors are safe to assume that the advice and information in this book are believed to be true and accurate at the date of publication. Neither the publisher nor the authors or the editors give a warranty, expressed or implied, with respect to the material contained herein or for any errors or omissions that may have been made. The publisher remains neutral with regard to jurisdictional claims in published maps and institutional affiliations.

This Springer imprint is published by the registered company Springer Nature Switzerland AG
The registered company address is: Gewerbestrasse 11, 6330 Cham, Switzerland

If disposing of this product, please recycle the paper.

*To Kelly, James, and Elizabeth,
of course.*

Acknowledgements

The research going into this book was inspired by much of the work done as part of the John Templeton Foundation grant "Agency, Directionality, and Function," and I would like to thank Alan Love for a very productive and helpful grant leadership. Chapters 2, 3, and 4 substantially expand and revise parts of my "What's the Point? A Presentist Social Functionalist Account of Institutional Purpose." *Philosophy of the Social Sciences* (special issue on the "Philosophy of the Social Sciences Roundtable"), 2022, 52(1–2): 53-80. Chapter 5 is a revised version of my "Institutional Corruption: The Teleological and Non-Normative Account." *Journal of Ethics and Social Philosophy*, 2023, 25: 464–494. Parts of Chap. 6 draw on Clint Hurshman's "Artifacts and intervention: A social account of artifact functions," *Synthese*, forthcoming.

Contents

1	**The Theoretical Backbone—Social Institutions, Functionalism, and Social Science**	1
	1.1 Introduction	1
	1.2 Social Institutions	2
	1.3 The Function of Social Institutions: Some Historical Remarks	5
	1.4 A New Social Science of Institutional Purpose: The Methodology of the Book	6
	1.5 The Structure of the Book	8
	References	9
2	**Missing Mechanisms, Arbitrary Assignments, and Counterfactual Conundrums—Existing Accounts of Social Functionalism and Their Problems**	13
	2.1 Introduction	13
	2.2 Historical Social Functionalism and the Missing Mechanisms Argument	15
	2.3 Structural-Functionalism and Its Problems	23
	2.4 Virtual Social Functionalism and Counterfactual Challenges	26
	2.5 The Intentional Design Based Account and Its Limitations	29
	2.6 Desiderata for a Compelling Account of Social Functionalism	30
	2.6.1 The Account Should Be Focused on the Present	30
	2.6.2 The Account Should Be Non-arbitrary	30
	2.6.3 The Account Should Be Actualist, Not Counterfactual	30
	2.6.4 The Account Should Be General	31
	References	31
3	**Presentist Social Functionalism—The Foundations**	35
	3.1 Introduction	35
	3.2 Presentist Social Functionalism	36
	3.2.1 The Account Should Be Focused on the Present	39
	3.2.2 The Account Should Be Non-arbitrary	40

		3.2.3 The Account Should Be Actualist, Not Counterfactual	41
		3.2.4 The Account Should Be General	41
	3.3	Objections and Responses	42
		3.3.1 General Objections	43
		3.3.2 Specific Objections	47
	3.4	Upshot	50
	References		51

4 Presentist Social Functionalism and the Function of Corporations . . . 55
   4.1 Introduction . . . 55
   4.2 The Function of Corporations . . . 56
   4.3 Presentist Social Functionalism and the Function of Corporations . . . 60
   4.4 Benefits of the Presentist Social Functionalist Analysis of Corporations . . . 66
   4.5 Conclusion . . . 70
   References . . . 70

5 Institutional Corruption: The Presentist Social Functionalist Account . . . 73
   5.1 Introduction . . . 73
   5.2 Institutional Corruption . . . 75
   5.3 Institutional Corruption: A Presentist Social Functionalist Account . . . 81
   5.4 Implications and Further Developments . . . 85
   5.5 Conclusion . . . 87
   References . . . 88

6 Artifacts: A Presentist Social Functionalist Account (Co-written with Clint Hurshman) . . . 91
   6.1 Introduction . . . 91
   6.2 Existing Accounts and their Problems . . . 92
   6.3 The Presentist Social Functionalist Account . . . 98
      6.3.1 Multiple Realizability . . . 101
      6.3.2 Multiple Utilizability . . . 101
      6.3.3 Recycling . . . 102
      6.3.4 Reproduction with Variation . . . 103
      6.3.5 Malfunction . . . 104
      6.3.6 Phantom Functions . . . 104
   6.4 Two Case Studies . . . 105
      6.4.1 Pharmaceuticals and off-Label Usage . . . 105
      6.4.2 ChatGPT and Other Large Language Models . . . 107
   6.5 Conclusion . . . 108
   References . . . 109

7 Conclusion . . . 111

Chapter 1
The Theoretical Backbone—Social Institutions, Functionalism, and Social Science

Abstract The chapter introduces the discussion of social functionalism. It characterizes the notion of a "social institution," which is foundational for the rest of the book, and sketches aspects of the history of the debate surrounding social functionalism. It also presents the methodological framework for the rest of book, and provides an overview of the latter's structure.

1.1 Introduction

Humans are social beings of a peculiar sort. It is not just the case that we live in groups of conspecifics—many social animals, including honeybees, chipmunks, and ducks, do that too. It is also not just the case that we depend on others' help and support to navigate the world around us—this is again something that is shared by many other organisms, from slime molds to chimpanzees. What makes us humans peculiar social animals is that our interpersonal relations are psychologically and normatively structured—and that in a dynamic manner.[1] We make our way through world by keeping track of what others are thinking (what their goals and motivations are, what they are feeling, and what they believe to be the case about their environment) and by what their place is in society (what they ought to do, given the social roles that they occupy) (Whiten and Byrne 1997; Sterelny 2003, 2012, 2021; Tomasello 2021, 2022). In short: we humans are peculiar for being psychological and normative social beings.

For present purposes, it is especially the latter part of this that is central: the fact that we live in normatively structured social environments. While the psychological and normative parts of human social living interact in many complex ways, they still

---

[1] Note also that *even if* understanding human social living were just a matter of analyzing human *behavior*—i.e. why A did X when B did Y—thus would be hard already. The complexities involved in behavioral ecology and sociobiology show that this difficult for many organisms. Humans are unlikely to be an exception here.

deserve to be given separate treatments (for more on the psychological side of human social living, see e.g. Sterelny 2003; Spaulding 2018; Gopnik 1996; Spelke 2022; Schulz 2025). In this spirit, the focus of this book is specifically on the fact that we humans live in normatively structured social environments. Of course, this does not mean that the psychological aspects of distinctively human sociality will be completely ignored (see e.g. Chap. 6), but the focus will be on the "external" aspects of human social living, not the "internal" psychological ones.

Now, there can be many sources of human social normativity, from moral imperatives and religious commandments to the demands of tool manufacture (Nichols 2004; Stanford 2018; Kumar and Campbell 2022; Birch 2021). However, for present purposes, it is not necessary to distinguish these sources of social normativity further: what matters here is just *that* there are these social commandments—norms—that prescribe how we are meant to act in a different circumstances. It is furthermore the case that these norms are widely thought to be central to the existence of *social institutions*. Since this notion is fundamental to the rest of the book, it is important to pause here and consider the nature of social institutions in more detail.

1.2 Social Institutions

There are three main conceptions of social institutions that have been defended in the social scientific and philosophical literature (see e.g. Guala 2016; Guala and Hindriks 2015; Bicchieri 2006; Miller 2010; North 1990).[2] While a detailed discussion of all of these is beyond the scope of this book, a few clarificatory remarks are in order here to place the rest of the book on a solid foundation. (Indeed, the rest of the book can also be seen as an extended argument for the power of the common, rule-based view of social institutions—though much of what follows can, if suitably adjusted, also be made to work with at least some of the alternative approaches.)

On the theoretically most basic approach towards social institutions, they are just *behavioral patterns in equilibrium* (Lewis 1969; Guala and Hindriks 2015; Guala 2016; Bicchieri 2006; Schotter 2008). As agents engage in social interactions, the outcomes of these interactions depend on what all the agents are doing. Now, as is well-known, these kinds of strategic interactions are in equilibrium when the players' actions are best responses to each other: given what the other players are doing, no player has an incentive to deviate from their chosen strategy. In the present context, this matters, as these equilibria can then be seen as *social institutions*. So, there is a social institution to drive on the left in Australia and on the right in the US, as, given that everyone else in Australia drives on the left, it is my best response to also drive on the left (and vice-versa for the US).

---

[2]This concerns the social scientific nature of social institutions; for more on the metaphysics of social institutions, see Witt (2023); Ruben (1985).

1.2 Social Institutions

While there many compelling aspects of this approach towards social institutions, the main flaw that it faces is that it is overly thin. On this approach, *all* equilibria turn out to be social institutions, and all social institutions turn out to be equilibria. However, from a social scientific point of view, we often want to distinguish these. For example, if I am an engaged in a one-shot interaction with just one other person—say, we are walking towards each other in the park and are trying not to hit each other—there is no need to invoke a highly local, one-time social institution that governs the behavior of the two of us that day in this specific spot in the park. On the other hand, there is reason to invoke social institutions even in case where agents are not engaged in strategic games: for example, we may want to appeal to the existence of a social institution when noticing that a significant number of the people attending a heavy metal concert are wearing black leather clothes—even though we may also note that these clothing choices are not best responses to each other (say, because choosing which clothes to wear is not a strategic interaction). Hence, this approach towards social institutions will not be central in what follows.

On the other extreme is the teleological approach towards social institutions: this one is theoretically extremely rich. On this conception, a social institution is (typically though not necessarily) an organization—a structured set of social roles—dedicated to achieving a collective end (perhaps in a specific manner), where this structure is underwritten by social norms (Miller 2010, 2017).[3] This is called a "teleological" approach towards social institutions, as it builds the *function* of a social institution into the characterization of its *nature*.

Again, there is much to recommend in this account—specifically, the appeal to social norms avoids the overly broad implications of the equilibrium account. Still, the account also faces some major difficulties; unsurprisingly, these are the flipside of the equilibrium account. In particular, the account can be harder to apply and work with, as it requires a very rich philosophical tapestry to even get off the ground. To determine whether something is a social institution, and if so, of what kind, we need to determine its structure, roles, collective ends, actions, and functions. These are all contested philosophical notions—thus making progress in the social sciences dependent on the resolution of these philosophical questions. However, in practice, it seems that there is more agreement on what social institutions there are—from social media to golf clubs—than on the identification of the appropriate collective ends, roles, activities, and functions.[4]

In between these two accounts is the most common characterization of social institutions: the one that sees social institutions as rules that structure human interactions and which set out the kinds of behaviors that, in a given type of situation, members of the society are expected to—and expect others to—engage in (see also

---

[3] Miller (2017) employs a morally loaded notion of social institution that is furthermore restricted to *organizations* (roughly, complex structures of organized sets of norms). However, as also noted in the text, this is not the standard notion used in the social scientific literature. For more on Miller's account, see Chap. 5 below.

[4] This is especially so for Miller's richer morally-loaded account: see also Chap. 5 for more on this.

Parsons 1951; North 1990; Searle 1995). Social institutions, in this standard social scientific sense, comprise a vast array of familiar aspects of contemporary social living, from the structure of the government (e.g., representative democracy) and the economy (e.g., free enterprise) to that of the family (e.g., polyandry) and religion (e.g., Hinduism). Note that it need not be obvious *why* social institution N prescribes behavior B in situation S. (This will become crucially important again momentarily.) Similarly, it is not presumed that the behavior prescribed by the institution is morally obligatory: institutionally-based norms are not necessarily moral norms.

The benefit of this account is that it avoids the pitfalls of either of the previous two types of accounts: it goes beyond seeing institutions as mere behavioral equilibria, but it is not so rich as to require the application of a major philosophical framework. For this reason, this is the account that will be the central in what follows.[5] While there is of course more that could be said about this account, this is all that is needed here—the rest of the book can be seen as an extended attempt to spell out its promises further. In short: for present purposes, it is key to note that social institutions should be seen to dictate the norms of behavior for a given society. Hence, making sense of how and why people act in the ways that they do thus depends on understanding these institutions.[6]

---

[5] Note that this account can be combined with the equilibrium account to see social institutions as (embodied) rules for equilibrium selection (Guala 2016; Guala and Hindriks 2015). However, to what extent this kind of combination is either needed or avoids the problems of the teleological account is not clear. At any rate, this will not be considered further here; for what follows, it is possible to accept the rules-in-equilibria account of social institutions, if that is so desired.

[6] It is important to acknowledge, again, that the appeal to social institutions is not all there is to making sense of human social behavior. In particular, since institutional norms need to be juggled and integrated—a person can be a volunteer firefighter, a professional podcaster, and a singer in a 1980's pop rock band at the same time, with different norms of behavior being entailed by the different institutions (Hollis 1994, Chap. 8)—we also need to consider how individual people think about their place in society (Munger 2019). On top of that, as just noted, people also navigate their social environments by considering the mental states of others (Nichols and Stich 2003; Goldman 2006): they do just think about the relevant social institutions, but also about what others want and believe (independently of the social institutions they or the focal agent are part of). Finally, and not unrelatedly, we could also think about what the ideally appropriate—rational—ways of acting would be in the case at hand, and, by assuming that people are generally rational, infer that they in fact do that (Hausman 2012; Rosenberg 2012; Satz and Ferejohn 1994). The fact that other theoretical tools are also important in the social sciences does not reduce the importance of the appeal to social institutions: the latter may not be able to answer all questions, but it can still be important for answering some of these questions. A version of this point will also become important for social functionalism in the later parts of this book.

1.3 The Function of Social Institutions: Some Historical Remarks

Given their importance to human thought and action, it is unsurprising that much work in the social sciences is centered on social institutions (see e.g. Guala 2016). For example, there is much work done on which political and economic institutions are conducive to high economic growth and stable societies (see e.g. Acemoglu et al. 2001; Robinson et al. 2005) and on which religious institutions underpin a liberalist free market society (Henrich 2020; Weber 1958).

However, this does not mean that an appeal to social institutions is necessarily easy. Which social institutions are there in a given society? Why are *these* institutions part of this society? Which of these institutions are key parts of this society, and which are peripheral ones that come and go out of existence quickly? Why do the institutions prescribe the behaviors that they do? It is not obvious how to answer these questions. Given the importance of the appeal to social institutions in social analysis, though, doing so is very important: to really understand why people do what they do, it is important to understand why the institutions in their society are the way that they are.[7]

One major way of answering these questions is by appeal to the *function* of social institutions: we want to know what social institutions are *for*. Unsurprisingly, therefore, the appeal to the function of social institutions is about as old as the social sciences themselves, and continues to be very important (see e.g. Durkheim 1915 [1971]; Malinowski 1922; Elster 1979; Bigelow 1998; Pettit 1996; Booth 2003; Holmwood 2005; Potts et al. 2016). So, for example, functionalist perspectives have been a key element of the analysis of rituals: such rituals—e.g. harvest festivals—have been seen as being *for* increasing the cohesion of the society by integrating its members and relieving internal tensions (Eisenstadt 1990; Fallding 1963).

This, though, does not mean that this appeal is not also very controversial. In particular, assigning functions to social institutions is not straightforward. In one of the earliest treatments of this issue, Durkheim (1915 [1971]) saw societies as akin to superorganisms, and the function of the social institutions as whatever features contribute to this organism being "healthy" and persisting over time. While Durkheim's own relationship with explicitly "functionalist" analysis was complex, his treatment was historically extremely influential in the social sciences (Pope 1975). Unsurprisingly, therefore, the literature on Durkheimian social analysis is large; fortunately, a detailed treatment is not necessary here. For present purposes, it is sufficient to note two points.

---

[7]A quick terminological point. The term "understanding" is often used in the social sciences to signal a particular hermeneutical approach towards the subject (Weber 1958; Rosenberg 2012; Hollis 1994). However, this is not how the term is used here; here, the term is used in a more general, explanation-focused sense—though no commitment is made that explanation always requires understanding, or exactly what understanding amounts to (see also Elgin 1991; Grimm 2010; Khalifa 2012).

First, there is no question that the view has some questionable metaphysical presuppositions—in particular, the ideas that societies are *actually* superorganisms or that we can explain causes by their effects. This has led to a backlash among social scientists and philosophers (especially logical positivists) (Rosenberg 2012; Hempel and Oppenheim 1948). On the other hand, though, aspects of the view have been refined and taken up in the literature; key among these is the idea that the function of social institutions should be seen as their contribution to the persistence of a wider social system. Importantly, this need not require that effects explain their causes—though it does require some kind of feedback mechanism. This is an important point to which I return in in more detail in Chaps. 2 and 3. Interestingly, also, the idea that social functionalism can profit from adopting insights from biology is important—and something that is a key part of the account defended in this book (see Chap. 3). However, it remains the case that, while Durkheim's views have laid the foundation for functionalist analysis in the social sciences, they are not the final formulation of this kind of analysis.

Indeed, several other accounts of social functionalism have been formulated and have come to be very influential in their own right. While discussing these accounts in more detail is the aim of Chap. 2, it is useful to note here that these alternative accounts have also been shown to face some challenges. For example, appealing to the history of a social institution is often tricky, as many social institutions lack such a history (Elster 1979; Kincaid 1990; Pettit 1996; Bigelow 1998). More generally, it is just not clear exactly what it implies when it is said that *social institution N has function F*. Despite much effort in addressing this issue, many open questions here are still remaining.[8]

This is important to note here, as this book seeks to make progress on exactly this point. It provides a novel answer to the question of what the function of a social institution is and goes on to justify this answer as being an improvement over the major existing treatments. However, from the get-go, it is important to understand three aspects of the nature of the project of this book.

1.4 A New Social Science of Institutional Purpose: The Methodology of the Book

What follows in the rest of the book is a philosophical, principled look at the theory of social functionalism. The goal is not to place the different versions of social functionalism into their historical context, follow their development over time, or consider how the different forms of social functionalism have historically responded to each other (for an analysis like that, see e.g. Bigelow 1998; Holmwood 2005). Rather, the goal is to look at archetypal versions of these theories and consider their

---

[8]This has been a classic concern with functionalist analysis in the social sciences: see e.g. Davis (1959).

structural strengths and weakness. The goal thus an assessment of the merits of the theories as such.

Second, however, the account of functionalism defended here is still focused on the social realm. There is a lot of philosophical concerning functionalism in general—concerning applications in biology, cognitive science, and medicine, to name a few (see e.g. Allen et al. 1998; Ariew et al. 2002; Cummins 1975; Garson 2012; Lombrozo and Carey 2006; Piccinini and Garson 2014). While, as the rest of the book makes clear, much of this work is an important influence on and building block for the account here laid out, the goal in what follows is restricted to questions of the functions of social institutions specifically. Similarly, while the arguments to follow may also have implications for functional ascriptions in other contexts—e.g. in biology or cognitive science—spelling this out is not part of the present project.

Third, it is important to emphasize that the book takes a broadly "scientific" approach towards the question of the function of the social institutions. There is no question that this is a controversial decision; indeed, many scholars see no value in such an approach, and instead favor a "hermeneutical" approach that emphasizes a personal, historicist, narrative, understanding-focused take on social questions (see e.g. Agar 2021; Taylor 1971; Giddens 1984). On the latter perspective, social analysis should be centered on—among other things—how the participants of a social institution understand the institutions they are part: what these institutions mean for and to them. Trying to ground social analysis on the functions of social institutions—especially where these function are spelled out in non-meaning-based, "objective" ways, as is done here—is thus the very opposite of this hermeneutical approach.

Now, given the long and very contentious history of the "Methodenstreit" in its many guises, there is little hope that there is an argument that could be presented in this chapter or the book as a whole that would settle this dispute in ways that would satisfy defenders of the hermeneutical approach. However, what *is* possible is to at least make the disagreement explicit and thus to make the theoretical choice points clear. Indeed, doing this is the aim of the rest of the book: the book aims to *show* what the social scientific perspective on social functionalism can do. That is, the book as a whole tries to underwrite the contention that it is possible to formulate a compelling, scientifically-grounded view of social function. While this may not convince those who have fully made up their mind about these issues, it should be of relevance to those approaching these questions at least with an open mind that allows for the possibility that a scientifically-grounded, empiricist view of social functionalism may have something to offer to the social sciences.

Fourth and finally, though, while the defense of presentist social functionalism is principled-philosophical and not sociological-historical, this does not imply that it is constrained to highly abstract considerations only. On the contrary, the book is a deeply application-focused treatment of social functionalism. In fact, this application focus is a two-way street. On the one hand, by focusing the critical assessment and defense of social functionalism—in the new version laid out in what follows—on a few specific sets of cases (the function of corporations, system corruption, and the function of artefacts), an overly abstract treatment of this new version of functionalism is avoided. The strengths and weaknesses of the different forms of social functionalism come out best in the context of specific applications, and this book avails itself of this fact. On the other hand, it is also the case that the look at specific applications points strengths and weaknesses of different accounts of social functionalism that would otherwise be overlooked. That is to say, it is not just the case that the applications illustrate the critical assessment of different views of social functionalism. Rather, it is also the case that these applications are *integral components* of this kind of assessment.

1.5 The Structure of the Book

With this in mind, the rest of the book is structured as follows. In Chap. 2, I present the main existing ways of spelling out the function of social institutions: (a) historical social functionalism, (b) structural functionalism, (c) Pettit's recent virtual social functionalism, and (d) intentionalist social functionalism. In each case, the chapter then presents, updates, and analyzes the most compelling objections to these traditional forms of social functionalism. The upshot of the chapter is the conclusion that a new approach to the topic of the function of social institutions is needed.

Chapter 3 then develops the new account of social functionalism that is at the heart of the book: presentist social functionalism. In particular, the chapter argues that social functionalism should be based on the *actual* bio-cultural selective or sorting pressures on the institution *as it is now*. The chapter shows how this idea can be deepened and underwritten with recent insights from evolutionary biology and the philosophy of biology. It also considers a number of the key objections to this kind of view that have been or could be proposed, and shows why they fail to be compelling. In this way, the chapter lays the theoretical and argumentative ground for the further, application-focused development of presentist social functionalism.

Chapter 4 begins this further development and defense of presentist social functionalism by considering a widely discussed question about an important contemporary social institution: namely, the question of the function of corporations. As is well known, two main answers to the question of what corporations are for have been proposed in the literature: the shareholder-value theory and the stakeholder-value theory. The chapter lays out these two views, and then argues that an appeal to presentist social functionalism shows that the terms of this debate are too constrained. In particular, by applying presentist social functionalism to this debate, it becomes clear that there are further possible functions of corporations that have not even been considered in the literature, and that existing studies of this question have looked at the wrong data, and thus fail to be able to resolve it. This chapter thus shows the practical applicability and fruitfulness of presentist social functionalism—especially compared to its rivals.

Chapter 5 considers systemic corruption. Corruption is widely recognized to be a major social problem, but its characterization continues to be very controversial. Indeed, it is now commonly noted that what is being corrupted need not be an individual person at all but can be an entire social institution. This kind of institutional corruption has, especially in the last few years, come to be seen as ever more central and important. In this chapter, I advocate for a novel theory of this phenomenon, according to which it is the result of an individual or collective agent acting in ways that prevent a social institution from partially or fully fulfilling its function. In turn, the function of a social institution is spelled out in line with presentist social functionalism. The chapter shows that this new theory of institutional corruption is a useful addition to the literature, as it situates the study of this phenomenon in a wider functionalist approach toward the social sciences and does justice to the complexity of institutional corruption—both when it comes to its inherent nature and its moral evaluation. In this way, this chapter shows how presentist social functionalism can be the basis for new developments in adjacent fields, and unify them with existing work in the social sciences.

Chapter 6 (co-written with Clint Hurshman) begins by noting that assigning functions to artifacts and technologies is crucially important for a number of different reasons: in particular, it can explain how and why artifacts are used in the ways that they are, which artifacts are stable parts of society. In turn, this can help justify social policies and interventions. However, how to assign functions to artifacts is not yet fully clear. The most popular attempt to do so appeals to the intentions of the designer of the artifact—though others have tried to do so by appeal to the history of use of the artifact. However, in this chapter, we show that neither of these theories is fully compelling, and that a better account of artifact function can obtained by applying a variant of presentist social functionalism to this case. We lay out this presentist theory of artifact functions and apply it to two examples of controversial artifact functions: off-label uses of pharmaceuticals, and policy responses to ChatGPT and other large-language models. In this way, the chapter expands the criticism of alternative theories of social functionalism by considering the intentionalist account in more detail, and makes the wide reach of presentist social functionalism clearer.

Chapter 7 summarizes the overall argument of the book, draws out some general themes, and points in the direction of further work on this topic. The chapter also emphasizes the methodological innovations of the book as a whole: in particular, it shows how the book brings together cutting-edge work in biological sciences with that in the social sciences in an application-focused context.

References

Acemoglu, D., S. Johnson, and J. A. Robinson. 2001. The Colonial Origins of Comparative Development: An Empirical Investigation. *The American Economic Review* 91 (5): 1369–1401.

Agar, M. 2021. *The Lively Science: Remodeling Human Social Research*. New York: Routledge.

Allen, C., M. Bekoff, and G. V. Lauder, eds. 1998. *Nature's Purposes: Analyses of Function and Design in Biology*. MIT Press.

Ariew, A., R. Cummins, and M. Perlman, eds. 2002. *Functions: New Essays in the Philosophy of Psychology and Biology*. Oxford University Press. https://books.google.com/books?id=ch0PiXfrhYwC.

Bicchieri, C. 2006. *The Grammar of Society: The Nature and Dynamics of Social Norms.* Cambridge: Cambridge University Press.
Bigelow, J. C. 1998. Functionalism in Social Science. In *Routledge Encyclopedia of Philosophy.* Taylor and Francis.
Birch, J. 2021. Toolmaking and the Evolution of Normative Cognition. *Biology and Philosophy* 36 (1): 1–26.
Booth, R. A. 2003. Form and Function in Business Organizations. *The Business Lawyer* 58 (4): 1433–1448.
Cummins, R. 1975. Functional Analysis. *The Journal of Philosophy* 72 (20): 741–765.
Davis, K. 1959. The Myth of Functional Analysis as a Special Method in Sociology and Anthropology. *American Sociological Review* 24 (6): 757–772. https://doi.org/10.2307/2088563.
Durkheim, E. 1915 [1971]. *The Elementary Forms of the Religious Life* (J. W. Swain, Trans.). London: Allen & Unwin.
Eisenstadt, S. N. 1990. Functional Analysis in Anthropology and Sociology: An Interpretative Essay. *Annual Review of Anthropology* 19:243–260. http://www.jstor.org/stable/2155965.
Elgin, C. Z. 1991. Understanding: Art and Science. *Midwest Studies in Philosophy* 16:196–208. https://doi.org/10.1111/j.1475-4975.1991.tb00239.x.
Elster, J. 1979. *Ulysses and the Sirens.* Cambridge: Cambridge University Press.
Fallding, H. 1963. Functional Analysis in Sociology. *American Sociological Review* 28 (1): 5–13. https://doi.org/10.2307/2090451.
Garson, J. 2012. Function, Selection, and Construction in the Brain. *Synthese* 189:451–481.
Giddens, A. 1984. *The Constitution of Society.* Berkeley: University of California Press.
Goldman, A. 2006. *Simulating Minds.* Oxford: Oxford University Press.
Gopnik, A. 1996. The Child as Scientist. *Philosophy of Science* 63 (4): 485–514.
Grimm, S. R. 2010. The Goal of Explanation. *Studies in History and Philosophy of Science Part A* 41 (4): 337–344.
Guala, F. 2016. *Understanding Institutions: The Science and Philosophy of Living Together.* Princeton: Princeton University Press.
Guala, F., and F. Hindriks. 2015. Institutions, Rules, and Equilibria: A Unified Theory. *Journal of Institutional Economics* 11 (3): 459–480.
Hausman, D. M. 2012. *Preference, Value, Choice, and Welfare.* Cambridge: Cambridge University Press.
Hempel, C. G., and P. Oppenheim. 1948. Studies in the Logic of Explanation. *Philosophy of Science* 15 (2): 135–175. http://www.jstor.org/stable/185169.
Henrich, J. 2020. *The WEIRDest People in the World.* New York: Farrar, Straus and Giroux.
Hollis, M. 1994. *The Philosophy of Social Science: An Introduction.* Cambridge: Cambridge University Press.
Holmwood, J. 2005. Functionalism and Its Critics. In *Modern Social Theory: An Introduction*, ed. A. Harrington, 87–109. Oxford University Press.
Khalifa, K. 2012. Inaugurating Understanding or Repackaging Explanation? *Philosophy of Science* 79 (1): 15–37.
Kincaid, H. 1990. Assessing Functional Explanations in the Social Sciences. *PSA: Proceedings of the Biennial Meeting of the Philosophy of Science Association* 1990:341–354. http://www.jstor.org/stable/192715.
Kumar, V., and R. Campbell. 2022. *A Better Ape.* Oxford: Oxford University Press.
Lewis, D. 1969. *Convention.* Cambridge, MA: Harvard University Press.
Lombrozo, T., and S. Carey. 2006. Functional Explanation and the Function of Explanation. *Cognition* 99 (2): 167–204. https://doi.org/10.1016/j.cognition.2004.12.009.
Malinowski, B. 1922. *Argonauts of the Western Pacific: An Account of Native Enterprises and Adventure in the Archipelagoes of Melanesian New Guinea.* London: Routledge.
Miller, S. 2010. *The Moral Foundations of Social Institutions.* Cambridge: Cambridge University Press.

References

Miller, S. 2017. *Institutional Corruption: A Study in Applied Philosophy*. Cambridge: Cambridge University Press.

Munger, M. 2019. *Is Capitalism Sustainable?* Great Barrington, MA: American Institute for Economic Research.

Nichols, S. 2004. *Sentimental Rules: On the Natural Foundations of Moral Judgment*. Oxford: Oxford University Press.

Nichols, S., and S. Stich. 2003. *Mindreading. An Integrated Account of Pretence, Self-Awareness, and Understanding Other Minds*. Vol. 114. Oxford University Press.

North, D. C. 1990. *Institutions, Institutional Change, and Economic Performance*. Cambridge: Cambridge University Press.

Parsons, T. 1951. *The Social System*. London: Routledge.

Pettit, P. 1996. Functional Explanation and Virtual Selection. *The British Journal for the Philosophy of Science* 47 (2): 291–302.

Piccinini, G., and J. Garson. 2014. Functions Must Be Performed at Appropriate Rates in Appropriate Situations. *British Journal for the Philosophy of Science* 65 (1): 1–20.

Pope, W. 1975. Durkheim as a Functionalist. *The Sociological Quarterly* 16 (3): 361–379. http://www.jstor.org/stable/4105747.

Potts, R., K. Vella, A. Dale, and N. Sipe. 2016. Exploring the Usefulness of Structural–Functional Approaches to Analyse Governance of Planning Systems. *Planning Theory* 15 (2): 162–189.

Robinson, J. A., D. Acemoglu, and S. Johnson. 2005. Institutions as a Fundamental Cause of Long-Run Growth. In *Handbook of Economic Growth 1A*, ed. P. Aghion and S. Durlauf, 386–472. North-Holland.

Rosenberg, A. 2012. *Philosophy of Social Science*. 4th ed. Boulder, CO: Westview Press.

Ruben, D.-H. 1985. *The Metaphysics of the Social World*. London: Routledge.

Satz, D., and J. Ferejohn. 1994. Rational Choice and Social Theory. *The Journal of Philosophy* 91 (2): 71–87.

Schotter, A. 2008. *The Economic Theory of Social Institutions*. Cambridge: Cambridge University Press.

Schulz, A. 2025. *It's Only Human: The Evolution of Distinctively Human Cognition*. New York: Oxford University Press.

Searle, J. 1995. *The Social Construction of Reality*. New York: Free Press.

Spaulding, S. 2018. *How We Understand Others: Philosophy and Social Cognition*. London: Routledge.

Spelke, E. 2022. *What Babies Know*. Oxford: Oxford University Press.

Stanford, P. K. 2018. The Difference Between Ice Cream and Nazis: Moral Externalization and the Evolution of Human Cooperation. *Behavioral and Brain Sciences* 41 (e95):

Sterelny, K. 2003. *Thought in a Hostile World: The Evolution of Human Cognition*. Oxford: Wiley-Blackwell.

Sterelny, K. 2012. *The Evolved Apprentice: How Evolution Made Humans Unique*. Cambridge, MA: MIT Press.

Sterelny, K. 2021. *The Pleistocene Social Contract: Culture and Cooperation in Human Evolution*. Oxford: Oxford University Press.

Taylor, C. 1971. Interpretation and the Sciences of Man. *The Review of Metaphysics* 25 (1): 3–51.

Tomasello, M. 2021. *Becoming Human: A Theory of Ontogeny*. Cambridge, MA: Harvard University Press.

Tomasello, M. 2022. *The Evolution of Agency*. Cambridge, MA: MIT Press.

Weber, M. 1958. *The Protestant Ethic and the Spirit of Capitalism*. New York: Scribner.

Whiten, A., and R. W. Byrne, eds. 1997. *Machiavellian Intelligence II: Extensions and Evaluations*. Cambridge University Press.

Witt, C. 2023. *Social Goodness: The Ontology of Social Norms*. New York: Oxford University Press.

Open Access This chapter is licensed under the terms of the Creative Commons Attribution 4.0 International License (http://creativecommons.org/licenses/by/4.0/), which permits use, sharing, adaptation, distribution and reproduction in any medium or format, as long as you give appropriate credit to the original author(s) and the source, provide a link to the Creative Commons license and indicate if changes were made.

The images or other third party material in this chapter are included in the chapter's Creative Commons license, unless indicated otherwise in a credit line to the material. If material is not included in the chapter's Creative Commons license and your intended use is not permitted by statutory regulation or exceeds the permitted use, you will need to obtain permission directly from the copyright holder.

Chapter 2
Missing Mechanisms, Arbitrary Assignments, and Counterfactual Conundrums—Existing Accounts of Social Functionalism and Their Problems

Abstract This chapter presents the main existing ways of spelling out the function of social institutions: (a) historical social functionalism, (b) structural functionalism, (c) Pettit's recent virtual social functionalism, and (d) intentionalist social functionalism. In each case, the chapter then presents, updates, and analyzes the most compelling objections to these traditional forms of social functionalism. The upshot of the chapter is the conclusion that a new approach to the topic of the function of social institutions is needed.

2.1 Introduction

Understanding the function of social institutions is important both from the perspective of conducting social science (as noted in the previous chapter) and when it comes to the diagnosis and alleviation of major contemporary social problems. In particular, many of these problems appear to center on the extent to which social institutions do or do not function as they are meant to do (Thompson 1995, 2018; Lessig 2013; Miller 2017; Den Nieuwenboer and Kaptein 2008; Treviño et al. 2014). For example, several of the institutions meant to provide important social goods in many contemporary societies—such as the ability of every citizen to be equally able to influence political decision making, to be well informed about what is going on in the wider society, or to be secure—often seem to fail in their task. Newspapers around the world are full of reports of voting being made difficult for certain groups of people, of factual reporting being deliberately intertwined with opinion stating, or of security being unevenly enforced across a society (Thompson 1995; Lessig 2013; Miller 2017; Satz 2013). Determining the functions of social institutions can provide a fulcrum with which to understand, evaluate, and respond to this these issues: knowing more about what voting, the media, and policing are *for* can tell us something about what is wrong with some contemporary situations, and point us in the direction of rectifying these issues.

© The Author(s) 2025
A. W. Schulz, *Presentist Social Functionalism: Bringing Contemporary Evolutionary Biology to the Social Sciences*, Foundations for Interdisciplinarity in the Life Sciences: Concise Monographs,
https://doi.org/10.1007/978-3-031-94833-6_2

However, while it therefore seems clear that determining what social institutions are for is important both for social scientific and social policy-related reasons, it is not clear exactly how to justify a particular functional ascription to a given social institution. Put differently: the fact that functional ascription is social scientifically and social politically important does not mean that it is clear what the most compelling theory is of the nature of social functions. Several different answers to this question have been proposed in the literature, but they also have seen much criticism and discussion (Pettit 1996; Bigelow 1998). In turn, this threatens to pull the rug out from all attempts to analyze social phenomena using the tools of functional analysis, with some scholars concluding that there is very little of value in them (see e.g. Elster 1979).

It is the goal of this chapter to make this worry more precise. The chapter will consider the four key theories of social functionalism in the literature to date: the historical selectionist account, structural functionalism, virtual selection-based functionalism, and intentionalist social functionalism. (As noted in the previous chapter, the teleological account of social institutions has an in-built theory of social functionalism. The present chapter, though, looks at *separate* accounts of social functionalism. For more on the teleological theory, see Chaps. 1 and 5.) In each of these cases, it will present the relevant theories and evaluate them in light of some of the major criticisms that have been raised against them. The latter is important also due to the fact that recent work in evolutionary biology, anthropology, and philosophy of biology have sharpened the foundations of some of these criticisms considerably, requiring a novel look at them. While, as will become clearer, it remains true that existing work on this topic leaves many questions unanswered, it also becomes clearer (a) which questions, exactly are still left open, and (b) the kind of progress on this topic. In turn, these insights can then be used to formulate set of desiderata that a compelling account of social functionalism needs to satisfy. For this reason, the goal of this chapter is more constructive than destructive: the aim is to use the existing work on the underpinnings of social functionalism as a building block for the formulation of a novel, fully compelling form of social functionalism.

The chapter is structured as follows. Section 2.2 lays out the historical form of functionalism in the social sciences, updates and revises some of the key objections to it, and shows why it therefore cannot be considered fully compelling. In Sect. 2.3, I present an alternative form of social functionalism: structural functionalism, and show why it cannot be seen as making for a fully compelling theory of social functionalism either. In Sect. 2.4, I show why Pettit's recent attempt to overcome this objection also fails. In Sect. 2.5, I briefly consider the intentionalist form of social functionalism (though I postpone a more detailed discussion of it until Chaps. 4 and 6). In Sect. 2.6, I summarize the discussion and use it to formulate desiderata on a compelling account of social functionalism.

2.2 Historical Social Functionalism and the Missing Mechanisms Argument

On the standard account of social institutions, a given institution N (in a certain society) prescribes a set of behaviors B across a set of situations S (Parsons 1951; North 1990). Members of the society know that, to the extent that they are part of institution N, they ought to do B in S (at least ceteris paribus), and they expect others to know this, too. This knowledge and expectation then leads members of the society to often (though not always) do B in situation S, and to often (though not always) censure others that fail to do B in S. These general facts about social institutions raise several further questions, though.

In particular, one might wonder if there are some behaviors in B that are more central than others. For example, being a police officer might require responding to reports of a vehicle theft as well as directing tourists to local sites, but the latter may not seem a central element of the institution of the police force. Why is that though? What makes some behaviors B_1 through B_n (such as investigating reports of property theft and destruction) central parts of the institution of N, and others (such as providing information for tourists) not?

Similarly, we might wonder if social institution N is a stable part of this society, or if the behaviors it describes are mere fleeting elements that, like fashion trends, are unlikely to be repeated much. Is policing something that is a key building block of the society in question or is it something that can be eliminated (or defunded) without loss? What about corporate offices? Are these a passing fad, soon to be replaced with widespread working from home or on a gig basis? What about publicly funded, obligatory schooling? As noted above, answering these questions is not just important for purposes of predicting or explaining the structural features of a given society, but also for designing successful interventions.

Underneath these questions is the function of a social institution. To determine the nature and stability of social institutions, it can help know what these institutions are *for*. (Of course, this presumes that it is the case that social institutions *have* a purpose—at least often. However, it turns out that, as will be made clearer below, this presumption is quite reasonable.) As Pettit (1996, p. 300) notes:

> The tradition of thinking associated with the likes of Durkheim in the last century and Parsons in this is shot through with the desire to separate out the necessary and the reliable from the contingent and the ephemeral. The idea in every case is to look for the core features of a society and to distinguish them from the marginal and peripheral. Functionalist method is cast throughout the tradition as a means of providing "a basis—albeit an assumptive basis—for sorting out 'important' from unimportant social processes (Turner and Maryanski 1979, p. 135").

However, exactly what is it that grounds the function of a social institution? How can we determine what policing, voting, corporations, etc. are *for*?

Traditionally, the main way of answering these questions is constituted by an appeal to some form of bio-cultural evolution (Bigelow 1998; Rosenberg 2012;

Kincaid 1990; Elster 1979).[1] In the background of this idea is the fact that appealing to the selective history of the trait in question is one of the major ways of grounding functional ascription in the biological and cognitive sciences. So, on this view, the function of the human heart is to pump blood and not to make a certain kind of noise, because pumping blood is what the heart was selected for.[2] Humans with hearts that pumped blood (or whose hearts pumped blood more reliably or efficiently) had a greater expected reproductive success than those whose hearts did not pump blood (or as reliably or efficiently)—irrespective of the noise the hearts made. Hence, the fact that hearts pump blood (reliably or efficiently) supports the spread of hearts. This should thus be taken for their function.[3]

Several key traditional versions of functionalism in the social sciences propose to apply this same reasoning to the social realm (Elster 1979; Bigelow 1998; Kincaid 1990).[4] This yields the classic, historical form of social functionalism: a given social institution N has the function F if past tokens of N were biologically or culturally selected for to do F. That is, if those past tokens of N that did F had a higher chance to reproduce tokens of N than those tokens of N that did not, then N (now) has the function to do F.

A classic example of this kind of functionalism is Rappaport's (1968) analysis of ritualistic pig slaughter among the Tsembaga Maring of New Guinea. The Tsembaga Maring raise pigs in abandoned gardens and consume them during ritualistic slaughter festivals. According to Rappaport (1968), the function of the ritualistic slaughter is to avoid damaging the soil in the gardens so that planting crops becomes harder: while having a small number of pigs is helpful for the planting of crops (as the pigs clear the grounds in the gardens), too many pigs are damaging. Hence, the regular killing of pigs helps keep their number down, while still allowing them to have their beneficial effects of preparing the soil for planting (as well as providing protein for human consumption).[5]

However, this classic, historical form of social functionalism faces an equally classic objection: the so-called "missing mechanisms argument" (Elster 1979; Pettit

---

[1] As noted in the previous chapter, some older social scientific traditions—such as that of Durkheim (1915 [1971])—appealed to the mental states of some kind of collective mind. However, because of concerns with the metaphysical presuppositions of this kind of view, this approach is no longer central in the literature.

[2] Note that, in this way, this kind of account helps to address the so-called "disjunction problem" (a version of the first question about institutions above). A trait may have lots of features, but not all of these features—not everything that the trait does—is part of its function. Only those features that have been selected for are part of the function.

[3] Several different versions of such a view have been developed (Millikan 1984, 2002; Papineau 1987; Neander 2006; Garson 2012; Papineau and Garson 2019), but for present purposes, these differences do not matter; just the core idea of the account is important here.

[4] This thus shows that the biological, cognitive, and social sciences are closely connected—a key benefit of this view according to many of its defenders: see e.g. Durkheim (1915 [1971]).

[5] Rappaport (1979) goes beyond Rappaport (1968) in specifying what makes the pig slaughter among the Tsembaga Maring *ritualistic*, rather than merely regular. This is not so central here, though.

1996; Bigelow 1998). According to this argument, there is hardly ever any reason to suppose that actual social institutions have the kind of selective history to ground functional ascription in the above manner. While this is a familiar argument in the literature, it turns out that there are many more complexities here than are typically recognized—many of which are the result of the recent work in (the philosophy of) evolutionary biology. Noticing these complexities is important, as they help in the formulation of the desiderata of what a compelling account of social functionalism looks like.

The first point to note here is that, in its most common form, the "missing mechanisms argument" relies on the standard view of the nature of selective processes. According to this view, for a process of change concerning a particular set of entities to be genuinely a case of "natural selection," the following presuppositions need to be met (Godfrey-Smith 2009; Brandon 1990; see also Schulz 2020):

(a) There needs to be variation in the population of the entities in question.
(b) The entities in question need to be able to reproduce (Brandon 1990; but see also Godfrey-Smith 2009), and the offspring of the entities in question need to be able to (somehow) inherit features from their parents (though the exact nature of the inheritance processes—including any mutational elements—can be left open; see e.g. Boyd and Richerson 2005; Godfrey-Smith 2009; Sober 2014).[6]
(c) The features that are able to be inherited by the offspring of the entities in question need to drive differences in the reproductive success of these entities.

Now, as will be made clearer momentarily, there is room to question some or all of these conditions. However, it also needs to be noted that these conditions are widely accepted and that they have been foundational for a lot of work in the biological, cognitive and social sciences. Hence, these conditions are not an unreasonable foundation for the "missing mechanisms argument." This means a version of this argument built on conditions (a)–(c) cannot be brushed aside simply for being built on these conditions. This argument needs to be tackled head on.

At the core of the "missing mechanisms argument"—in this standard version—is the claim that, when it comes to most social institutions, conditions (a)–(c) are seldom if ever all met. First, many social institutions lack historical variation: they are not "tried out," in different versions, in one or several cultures. Rather, there was often only ever one such institution that was present. For example, in the case of ritualistic pig slaughter among the Tsembaga Maring, it is not like different rituals—some involving pig slaughter, some not—were tried out. Rather, there was ever only one kind of ritualistic slaughter.

---

[6] Hodgson and Knudsen (2010, pp. 94–104) use the labels "successor selection" and "subset selection" (derived from Price) for the distinction between reproduction and growth. Note also that, following most of the literature, no assumption is made here that selection requires *replication* in a strict sense (Godfrey-Smith 2009; Sober 1984, 2000). While Hodgson and Knudsen (2010) (e.g.) argue in favor of the need for replication, they employ a very broad sense of replication. This is thus more of a semantic than a substantive point.

Second, it is far from clear that social institutions reproduce, rather than merely *persist*. That is, it is often not clear (to say the least) that a given token social institution gives rise to one or more separate offspring token social institutions, which then go on to reproduce independently of their parent. Rather, this case may often be better described as simply seeing the social institution as *persisting*. So, it is not that different tokens of qwerty keyboards gave rise to more offspring qwerty keyboards compared to tokens of Dvorak keyboards and their offspring. Rather, this case seems better seen as involving the continued persistence of qwerty keyboards, and the gradual disappearance of the Dvorak keyboards. This matters, as genuine selection requires reproduction. If there is merely differential persistence, then this might qualify as a case of "sorting" (Schulz 2020; Vrba 1984)—but it would not be selection in the strict sense.

Third and finally, even where there is variation, even where this kind of selection exists, it is often purely extrinsic. That is, the features that determine which institutions persist or reproduce are chance-based. This makes the evolution more drift-like than selective. A familiar example of this is the adoption of the "qwerty" keyboard design, which was not adopted because of its inherent advantages compared to rival designs like the Dvorak keyboard, but because it happened to become popular early on, thus making it costly to switch away to alternative keyboard designs (David 1986; Lewin 2001).

In short: it is often thought that the traditional forms of social functionalism suffer from the fact that the kind of selection process they posit as grounding the function of social institutions did not in fact take place. Hence, traditional social functionalism is often thought to be empirically implausible.

Now, it needs to be noted that this classic criticism of traditional social functionalism may be somewhat overstated. It turns out that the bio-cultural selective history of a number of social institutions does satisfy assumptions (a)–(c) (see e.g. Boyd and Richerson 2005; Henrich and McElreath 2007; Wilson and Gowdy 2013). For example, different ways of pronouncing words (e.g. by dropping a leading "h") may well evolve by a process of selective copying, as they can provide individuals ways of identifying useful role models (Boyd and Richerson 2005). Similarly, some corporations may well evolve in a selective manner in competitive markets (Schulz 2020). Related points can be made about different forms of music, different moral frameworks, or different political systems (Henrich 2015; Heyes 2018; Nichols 2004).

However, what remains true is that there are many social institutions for which this is not the case (Pettit 1996; see also Schulz 2020). This is important, as for social functionalism to be a compelling and useful approach in the social sciences, it needs to be able to ground functional ascription in a wide set of cases. It may be defensible that some religious practices, some corporations, or some social mores have no function. However, functionalism as a research program loses much of its epistemic value if *most* religious practices, corporations, or social mores have no function. If only Dunkin Donuts has a function—as it has the right bio-cultural selective history—but Krispy Kreme, Starbucks, Walmart, and so on do not—as they lack this needed history—the functional analysis of corporations has little to add to the social sciences. For the latter, it needs to be the case that corporations

2.2 Historical Social Functionalism and the Missing Mechanisms Argument

typically have a function. Otherwise, the approach has too little methodological hold. The same goes for other social institutions (both types and tokens).

The upshot of this is therefore that the problem with the traditional, historically-focused version of social functionalism is not that it is never plausible. Rather, it is that it is not general enough. It may work in some cases, but as a general account of functional ascriptions in the social sciences, it cannot do the kind of work we ask it to do.

However, as noted earlier, all of this is premised on the fact that conditions (a)–(c) are accepted as grounding genuine selective processes. Recently, though, some philosophers and biologists have questioned the need for some or all of (a)–(c) as presuppositions of selective processes. For example, some authors have questioned the need for reproduction and allow even mere differential persistence as counting for selection—at least in some contexts (see e.g. Papineau and Garson 2019; Doolittle 2014; Schulz 2020; James et al. 2023; Price 1995; Fresco et al. 2018). The most radical version of this recent alternative approach towards selection is Doolittle's (2014) revised Gaia hypothesis.

According to Doolittle (2014), it is plausible to see selection occurring even in cases where there is only one entity that does not reproduce—such as the biosphere as a whole. The core idea behind Doolittle's argument is that, due to the second law of thermodynamics, we should expect any entity to sooner or later die out. The longer an entity persists, therefore, the more we have reason to think that the entity has features that prevent the extinction of that entity. Importantly, this is independent of how many entities there are in the population. In turn (Doolittle argues) this implies that there can be evolution without (a)–(c). Doolittle's argument rests on the idea that, in at least some contexts, there are many important similarities in these three scenarios (Figs. 2.1, 2.2 and 2.3):

In all of these cases, we see one entity—c—taking over the population over time. In scenario A, this happens because of traditional natural selection. In scenario B, this happens because of differential persistence. In scenario C, this happens because the entity in question successfully manages to stay in existence over the entire time period in question. While the underlying mechanisms are thus different here, the patterns these mechanisms give rise to appear quite similar.

Applied to the present case, this therefore means that we could see an institution N that was *not* tried out in different versions, and which does *not* genuinely replicate, as still having function F in much the same way as envisioned by the classic, historical form of social functionalism. We just need to read the claim that N was "selected for" doing F as the claim that N has feature set F, which caused it to have an increased chance of persistence, compared to what would have been the case purely by chance. For this reason, one might be inclined to say that the "missing mechanisms argument" fails to get off the ground here: there is no missing mechanism here at all—we just need to accept a broader set of mechanisms of selection.

However, as it turns out, this conclusion would be too quick. It is true that, especially when it comes to social institutions, it is not necessary to focus strictly on reproduction, and that differential persistence can also be seen as a source of a selection process in the relevant sense. This is an important insight to which I return

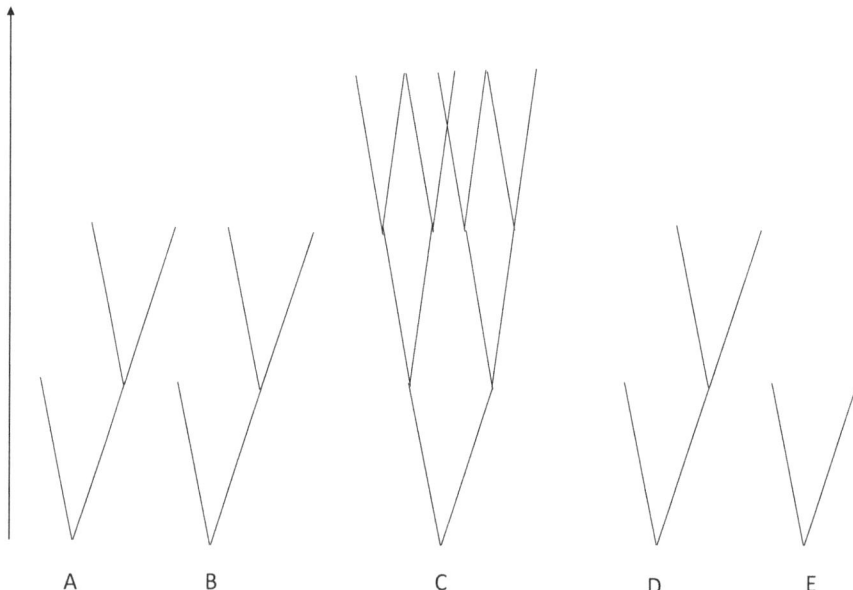

Fig. 2.1 Scenario A—standard natural selection

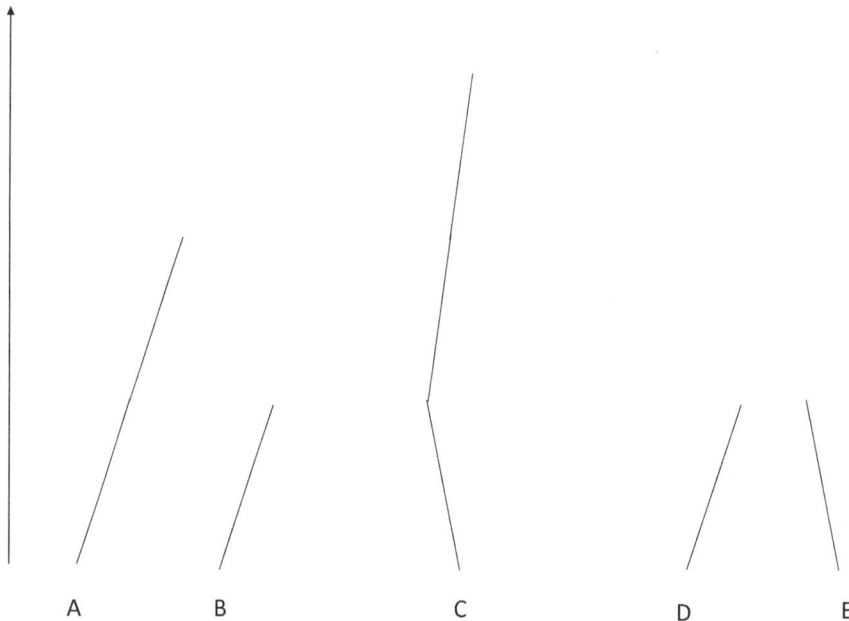

Fig. 2.2 Scenario B—differential persistence

Fig. 2.3 Scenario C—mere persistence

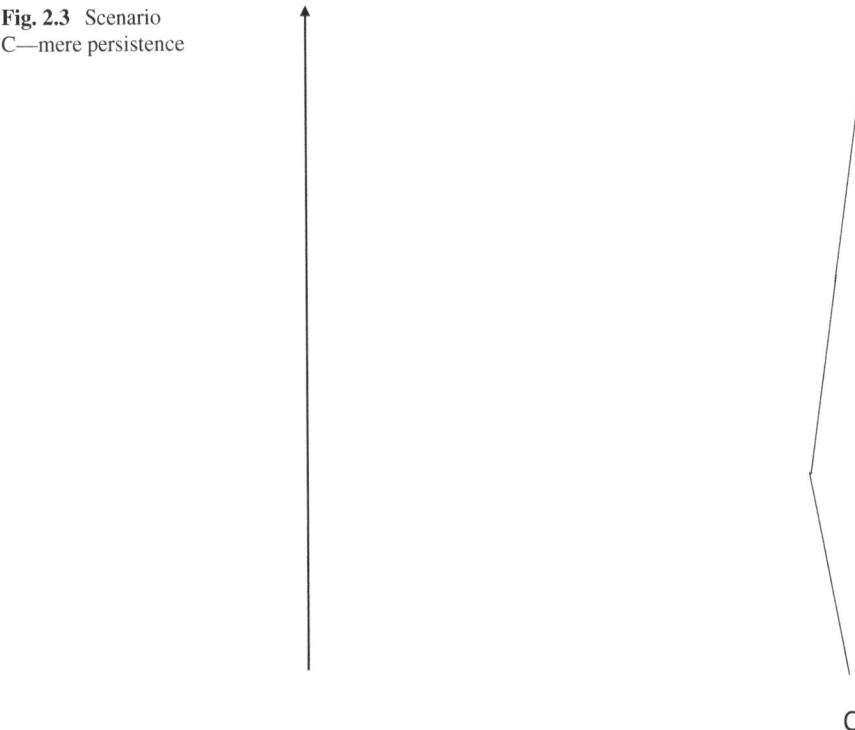

in Sect. 2.6 below, and it will also be a key element in the discussion of the next chapter. However, this does not mean that the classic, historical forms of social functionalism are now vindicated. This is not just because the other aspects of Doolittle's (2014) argument are quite controversial. In particular, there may well often be reasons to distinguish cases of differential reproduction with heritability from cases involving mere persistence, especially in a population of one (Godfrey-Smith 2009; Brandon 1990). However, even if we leave this worry aside, there are two further reasons why the appeal to Doolittle's alternative account of the nature of selection cannot, in fact, rescue the historical approaches towards social functionalism.

On the one hand, there is still a version of the extrinsicality worry here. In particular, Doolittle's (2014) argument for seeing single, merely persistent entities (like the biosphere) as being *selected* depends on the assumption that the relevant entities survive for a considerable period of time. Especially for shorter periods of time, random, drift-based persistence is entirely plausible. That is to say, it is true that the *longer* an entity E persists for, the *more plausible* it becomes to see E as having a feature G that increases its chances of survival. However, for short periods, this means that it is entirely plausible to see the persistence of E as having been mostly driven by chance. Importantly, exactly this is the situation of many social institutions: many social institutions—like the qwerty keyboard even ritualistic pig slaughter have only been around for very short periods. Indeed, given that human culture, as a whole, is relatively young, there is every reason to think that the persistence of many social institutions is heavily based on chance—or, at least, Doolittle's (2014) argument provides no reason to think otherwise. That argument may well apply to something like the persistence of our entire biosphere (which is also what it was designed to explain) but its applicability to social institutions is much less clear.

On the other hand, seeing the function of an institution as those of its features that increased its survival in the past is likely to yield a large and quite heterogenous set that is not particularly meaningful. Over time, social institutions may persist for many different reasons. The qwerty initially persisted because of successful marketing and then because it was already widely adopted. Ritualistic pig slaughter similarly may have persisted partly because it increased crop yields but also because children were taught that this is "the thing to do" at certain times, and then, once it got established, simply because it was embedded in the life of the community. More generally, it is likely that there is some set of features F_1 of an institution N that increased its chances of survival in some point in the past t_1, some other set of features F_2 that increased its chances of survival at t_2, and so on. We would then be forced to say that N has functions F_1 and F_2—even though they may be very different, and even though they may have nothing much to do with how the institution has recently operated.

Now, it may seem tempting to follow a common move in the literature on functions in the biological and cognitive sciences, and focus just on the *last* case of selection (Godfrey-Smith 2009; Piccinini and Garson 2014; Birch 2017). The trouble with this move is that this is still very ambiguous: since time is extended, it is not clear what the latest instance of selection is. If we focus our attention on the immediate past—last month, say—an institution may have survived for some reason R_1, but if we focus on the last five years, for some reason R_2, and so on. What is the "latest" instance of selection here? Note that this is particularly pressing here, since in the case of Doolittle's alternative account of selection, there are no generations and no reproduction, so selection is much less clearly defended than in the standard treatments. In short: it remains the case that it is not clear that appealing to the features that did increase an institution's chances of survival yields unambiguous assignments of functions. That said, the idea of appealing to the features of an institution that enhance an institution's chances of survival is important and is something to which I will come back. It is specifically the historical focus that is problematic here.

Because of these concerns, the classic, historical form of social functionalism still has to be seen to be problematic and subject to the "missing mechanism argument." Whether we see the relevant mechanism as a standard form of selection—with differential reproduction and heritability—or in a more heterodox manner—as merely involving persistence—it is not clear, at the very least, that most social institutions satisfy the conditions for the existence of the relevant mechanism, or that this would yield clear functional ascriptions. Hence, it is not clear that the classic, historical approach is a good way to ground the function of social institutions.

There are three alternative accounts of social functionalism that have been proposed that can avoid these challenges.[7] One appeals to the social structure of which a given institution is part, another to a firm of virtual selection, and a third to the intentions of a designer. However, as the next three sections make clear, none of these forms are fully compelling either.

2.3 Structural-Functionalism and Its Problems

The first way to avoid the issues with historical forms of social functionalism is based on appealing to facts about the structure a given social institution is part of. That is, some classic works in the structuralist functionalist tradition in the social sciences—such as that of Parsons (1951)—see social institutions as akin to elements in a large social system, and ground their functions in the causal roles they play in this larger social system. This is thus a social science-focused version of the more general causal role-based view of functional ascription in the biological and cognitive sciences (Cummins 1975), according to which functional ascription is analyzed in terms of the causal roles a trait or component plays in a larger causal system.[8]

So, for example, the function of a police force may be to safeguard law and order, as this is a key causal role it might play in a larger social system: the existence and behavior of the police force causes the safeguarding of law and order—which, in turn, contributes to the existence of social stability and therefore enables calm and deliberative political decision making, which is a key cause of the creation of a police force focusing on safeguarding law and order in the first place (Fig. 2.4).

Of course, this is not necessarily the only causal role the police force might play. If we focus our attention on the specifics of who is in power, the police force could also suppress dissent, thus maintaining the monopoly on power of a ruling elite—and thus keeping the police force on task to suppressing dissent (Fig. 2.5).

This sort of view is a foundational theory in many social sciences, has therefore seen much discussion and development (for overviews, see e.g. Vincent 2015; Potts et al. 2016; Bigelow 1998). However, for present purposes, it is sufficient to restrict the discussion to two key features of the view—one positive, and one negative. This is due to the fact that the goal here is not a detailed treatment of structuralist functionalism per se, but just to provide the motivation and foundation for the development of a new account of functionalism (to be laid out in the next chapter).

First, on the positive side, the view is "presentist:" it is focused on the present features of the institution in question. This is important, as it means that functional ascription does not need information beyond the social structure in question—in

---

[7] Note that this is not a historical statement: as noted in the text, structural-functionalism was developed at least somewhat independently of the classic, historical form of social functionalism.
[8] For a discussion of biological versions of this kind of view, see e.g. Novick (2023).

Fig. 2.4 A structuralist-functionalist analysis of policing

Fig. 2.5 An alternative structuralist-functionalist analysis of policing

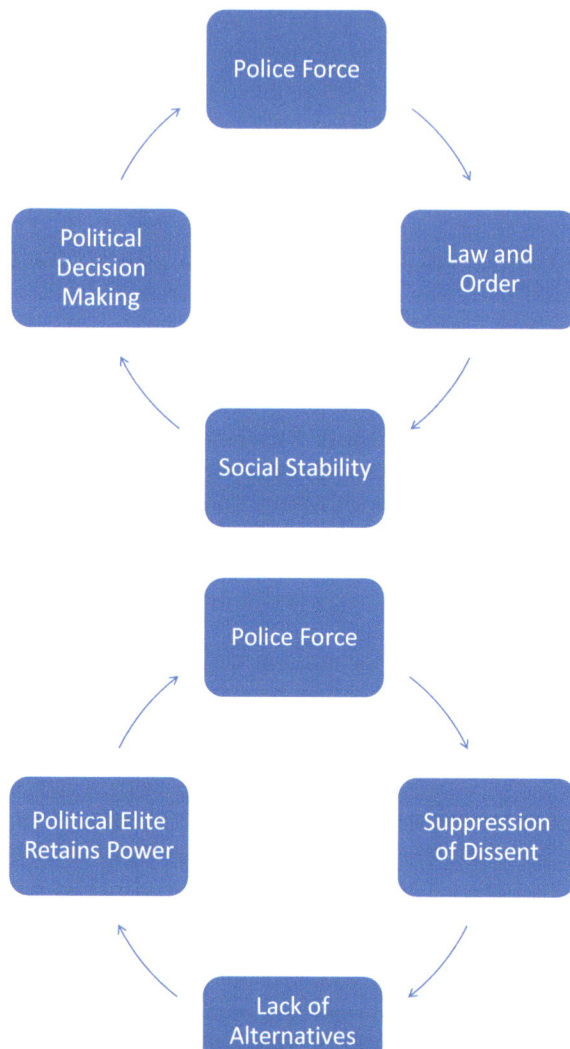

particular, the history of the institution or the social structure as a whole is not relevant. As just noted, this is a major benefit of the account that deserves to be stressed. In particular, it avoids the problems for the historical functionalist account just laid out. Indeed, it avoids some classic issues with historical analyses in general (see e.g. Turner 2007): historical processes—again in virtue of the second law of thermodynamics—tend to erase evidential traces of the past, making the justification of historical claims often very difficult. Of course, this is not to say that we can never find out anything about the past. Rather, the key point to note here is that by not making functional ascription dependent on historical claims, we are able to ascribe functions to institutions even in the absence of a detailed understanding of their history, and even if they do not have much history, as they have just been created. This is a major benefit of the account, as it enables it to have a much wider reach than historical forms of social functionalism. In turn, this gives this form of social functionalism a more compelling theoretical basis.

Second, though, on the negative side, this view is, in its nature, observer-dependent in a problematic manner: depending on what social system is chosen as the target of the analysis, the function of an institution will differ. Now, observer-dependence on its own is neither mysterious nor necessarily problematic. Observer interests determine what something is evidence for (whether a deck of cards is well shuffled or whether my hand is a winning hand) and observers can be part of a network of people that determine what something is (to be a 1-dollar bill is to be recognized as such). Here, though, we get a problematic form of observer-dependence: the same institution can be assigned different functions even by the same researcher and without changing the question that is asked (viz., what their function is). A dollar bill does not cease to be a dollar bill if a researcher chooses to conceptualize it as mere a piece of paper. However, exactly this is the case when it comes to the function of social institutions on the structural functionalist account.

This comes out clearly from the above examples: whether the function of the police force is the maintaining of law and order or the suppression of dissent depends on what overall system we are looking at. Indeed, this is a familiar worry with this kind of view (Kincaid 1990; Holmwood 2005; Vincent 2015; Ariew et al. 2002). Consider the human eye. On the one hand, when we focus on navigation and locomotion, its function could be taken to be to enable humans to *see*: after all, providing visual information is the key causal role eyes play in the human navigational-locomotive system. On the other hand, though, when we focus on social interaction, the function of the eye could be seen to consist in signaling mental states (intentions, say) to others: after all, a lot of mental state attribution depends on how and where someone looks, and given the importance of this to human social living (see e.g. Whiten and Byrne 1997; Spelke 2022), the function of the eye could be seen to rest in this.

This sort of observer-dependence is problematic, as it introduces an element of arbitrariness into the account. One of the main reasons why a functionalist approach has been attractive in the social sciences is precisely because it allows for an analysis of social dynamics that is non-arbitrary, but objective (Merton 1968; Pettit 1996; Kincaid 2009; Turner 2009b).[9]

One way to see the concern here is by noting that the view conflates *cause* and *function* (Millikan 1990; Papineau 1987). Lots of things can be caused by an institution or be the cause of it; however, we want to use functional ascriptions to single out some of these causes as being especially significant. For example, we may want to use functional ascriptions to predict and understand why certain institutions are

---

[9] Merton's legacy is a bit controversial, though: see e.g. Turner (2009a, b); Kincaid (2009); Campbell (1982) for discussion. Hayek and his followers could be seen to defend a non-historical form of functionalism as well (see also Kley 1994, Chap. 7).

more stable parts of society, and others less so (on which more in Chap. 4), not just what these institutions *do*. Similarly, we might want to use functional ascriptions to criticize certain actions as *corrupting* an institution (on which more in Chap. 5)— i.e. as preventing the institution from fulfilling its function, not just preventing it from doing something we happen to currently focus on. Finally, we might want to use functional ascriptions to pinpoint what an artifact *is* (on which more in Chap. 6)—not just what it is commonly used for, but what it *ought* to be used for. By contrast, it seems that the structuralist-functionalist account is forced to say that assignments of F being *the* function of an institution N is just a matter of our theoretical interest.

A different way of making this point is in terms of another common criticism of structuralist-functionalism: that it is inherently reactionary. The worry here is that since functional ascription is relativized to an existing social system, it is not clear how to make sense of subversive institutions (Rosenberg 2012). For example, how can we make sense of a political magazine, whose aim it is to disrupt the workings of the social system? If functional ascription is social system-relative, then it is not clear how it is possible to understand entities whose functions are subversive of that social system. Here, the structuralist-functionalist view faces the challenge that it seems to only be able to make sense of subversive institutions by focusing on the particular parts of the social system the latter are involved in. For example, the magazine above may be embedded in a causal-structural network involving specific political parties and voters, whose aims are to change the social system as a whole. In this way, the magazine turns out not to be subversive in and of itself; rather it is the causal structure that it is part of that is subversive of the larger social structure *it* is part of.

The trouble with this, though, is not just that it is oddly indirect—though it is that: it is not the magazine that is subversive, but the causal structure it helps to build up. More importantly, though, the trouble with this is that it *defines away* the possibility of inherently disruptive social institutions. Since the latter obtain their functional ascriptions due to the causal roles they play in maintaining some larger social structure, they cannot really go against this larger social structure.

For these reasons, while the structuralist-functionalist tradition certainly is an important social functionalist approach, it cannot be seen to be fully compelling either. Consider therefore the next non-historical form of social functionalism: the counterfactual one.

2.4 Virtual Social Functionalism and Counterfactual Challenges

A more recent, but philosophically very influential way to move social functionalist theorizing beyond the historically-focused versions is the account of Pettit (1996). While, as will be made clearer momentarily, this account faces some problems, it also has some key benefits: in particular, it has some important novel and interesting features.

2.4 Virtual Social Functionalism and Counterfactual Challenges

According to Pettit (1996), functional social institutions should not be seen as having a particular bio-cultural selective history—whether this is understood in the traditional or Doolittle's alternative sense—but as being *virtually* selected. In particular, for Pettit, for a social institution N to have the function F, it is not required that past tokens of N's having F led, through a process of differential replication, to present tokens of N having F; neither is it required that N's having F made it less likely for N to have gone extinct over the course of its existence. Rather, all that is required is that *if* the existence of N *were* threatened by some external factor, N's having F *would* ensure it continued to exist.

That is, according to Pettit (1996), social functionalism should be counterfactually grounded: what matters is not how some social institution actually arose. Rather, what matters is how the institution *would* respond if its continued survival *were* called into question. The source of the survival pressures on N can be manifold and include the arrival of an alternative social institution or other changes to the society in question. What is important is just that, on the account of Pettit (1996), N's having F means that it is due to N's having F that it is a stable part of the relevant society. N's having F ensures that N will continue to be part of the society even if its existence is otherwise threatened.

The example Pettit (1996) uses to illustrate this point is golf clubs. Assume, for the sake of the argument, that these clubs have not been bio-culturally selected for in the way required by the historically-focused form of social functionalism (in either its traditional or persistence-focused version). So, perhaps these clubs arose more or less accidentally when people regularly joined together at the same places to play the game and network. No alternative ways or arranging golf games or business networking opportunities were tried out, and the present clubs were therefore not the result of some selection or sorting among competing alternatives, and the short time of their existence makes the persistence-focused arguments non-applicable.

However, according to Pettit (1996), it may still be true that these clubs have the function to facilitate business relationships and networking opportunities. This would be so if these clubs would persist even if there were external shocks to their existence. So, assume further that it just so happened that people found it harder to learn to play golf (maybe changes in working hours limit the time available to spend on hobbies), or that the physical components needed to play golf such as the racquets or grassy plains became harder to obtain (e.g. because of climate change or materials shortages). If it then turns out that, even under these conditions, people would still join golf clubs so as to network with potential business partners (perhaps they would just take a directed, timed walk, or change the rules of the game in some way, such as just focusing on driving ranges), then golf clubs have the function to facilitate the latter, whatever may be true about the history of golf clubs.

Now, it needs to be noted that, in several respects, this response is a compelling way to handle the concerns of the traditional way of grounding functionalism in the social sciences. The step away from the historical focus of the traditional version of social functionalism is certainly moving in the right direction. As noted in the context of the discussion of structuralist-functionalism, moving away from a dependence on historical information enables functionalist theorizing to sidestep a number of worries. Similarly, the idea that the functions of social institutions should be seen as those of their features that make it more likely that they are part of future iterations of the relevant society is plausible, too: as noted in Sect. 2.1, this does get at a key motivation behind social functionalism. However, that said, the account of Pettit (1996) also faces several problems that prevent it from being fully compelling as it stands.

In the first place, the weight of the account rests entirely on the truth of the relevant counterfactuals. A social institution N has function F if N's having F ensures that N would continue to exist even if it had to face threads to its existence. However, it can be quite difficult to evaluate counterfactuals like this: how do we know what would happen if people had a hard time getting to golf clubs, or buying the necessary equipment, etc.? Would they still join the clubs? How do we know? Note that the point is not that we *never* know how to evaluate counterfactuals (Stalnaker 1968; Fodor 1990)—as also noted below, there are many contexts where we do have a good sense of what would happen if so and so were the case. The point is just that there are also many counterfactuals that we do not know how to evaluate, and many of these are crucial to the account of Pettit (1996). I may be able to say, for *some* worlds W, whether N would persist in W; however, for many other worlds W', I may not be able to determine this. (Would humans still play golf if they moved to Mars, or would they network differently? Given that social and technological living in the world where people live on Mars is bound to be quite different from what it is now, it is not clear how this question can be answered.) This matters, as it makes this account often difficult to apply. While we may sometimes be able to say if social institution N were to persist if so and so happened, we often would not be able to do so. In turn, this would leave it unclear whether N has a function—and if so, what that is.

Second and even more importantly, it is not clear which shocks social institution N is meant to be buffeted against. Is it all shocks? This seems overly strong: clearly, if a new social institution—urban hiking clubs, say—appeared that also facilitated business relationships and networking opportunities, but did so more easily than golf clubs, golf clubs might cease to exist. This, though, might not be seen to affect the fact that golf clubs have the function to facilitate business relationships. Indeed, it is precisely because urban hiking clubs coopt this feature that they can push golf clubs out of existence in this counterfactual scenario. However, if we are to limit the circumstances to consider when determining the function of N, how are we going to do this? On the face of it, it is not clear what a compelling answer to this question looks like. Without such an answer, though, functional ascription in the social sciences becomes arbitrary.

Third and finally, Pettit (1996) account is not fully spelled out. So, it is not made clear whether functional ascription really involves virtual *selection*—i.e. whether it is based on the counterfactual replication of the relevant social institutions—or merely virtual *sorting*—i.e. whether it is just based on the (counterfactual) growth or survival of the relevant social institutions. This is an important issue that, as made clear in Sect. 2.2 above, should be made explicit.

2.5 The Intentional Design Based Account and Its Limitations

The final account of functional ascription to be mentioned here is the intentional design-based one. It will be discussed in more detail in Chap. 4 and especially Chap. 6, but a brief statement of it is still useful here. According to the intentional design-based account of functional ascription, the function of something X is determined by the intentions of X's designer D (for versions of this kind of account, see e.g. Dipert 1993, 1995; Hilpinen 1993; McLaughlin 2000; Houkes and Vermaas 2004, 2010; Vermaas and Houkes 2003).[10] So, if D built X for reasons R, R is the function of X. For example, the function of a pair of scissors is to cut things with, since that is what they were designed to do. They can also be used (more or less well) as paper weights, crowbars, or musical instruments, but that is not their function, since that is not what they were designed for. We might say the same thing about social institutions: a social institution N has function F if we instituted N to achieve F. For example, we may argue that the function of public universities is to create an informed and intellectually nimble citizenry because that that's what they were designed to do. While universities also advance scientific knowledge or provide new technologies, that's incidental to their mission, and thus not part of their function.

As will become clearer in Chaps. 4 and 6, there may well be cases where the intentions of designer influence the function of something (Millikan 1984, 2002). The problem is that this is not *generally* the case. Not all institutions have a clear intentional designer: the institution of marriage, for example, was not clearly designed by anybody. This means that this account of functional ascription is inherently limited and does not speak to many cases of interest. This makes it unsuitably as a general approach towards social functionalism. Still, as made clearer in Chap. 6, it does deserve mention here, as it is a central treatment in some parts of the literature.

---

[10] As also noted in Chap. 1—and will be made clearer in Chap. 5—the teleological account of social institutions may also be seen as an intentionalist (though not necessarily) design-based account of institutions. Something similar goes for Searle's "collective acceptance account" (Searle 1995). As also noted earlier, though, these are not functionalist accounts, and so will not be further discussed here.

2.6 Desiderata for a Compelling Account of Social Functionalism

Pulling all the strains of the discussion thus far together, we obtain four desiderata for a compelling account of social functionalism.

2.6.1 The Account Should Be Focused on the Present

If the account of social functionalism is based on historical facts, it is unlikely to be able to ground functional ascriptions to social institutions in general. Social institutions tend to be too recently created to allow appeal to a purely persistence-based form of selection and lack the kinds of features—variation, heritability, and reproduction—to make appeal to standard reproduction-based forms of selection compelling. By appealing to features of the present only, compelling accounts of social functionalism avoid the problems that befall historical analyses in general, in that functional ascription is no longer held hostage to historical information that we may not have, or which may not obtain.

2.6.2 The Account Should Be Non-arbitrary

A compelling account of social functionalism does not make functional ascription *merely* a matter of our individual interests. Of course, which social systems we want to study, and which questions we want to ask about these social systems is a matter of our theoretical interests, as well as our moral, political, and aesthetic commitments (Potochnik 2017; Fox Keller and Longino 1996; Douglas 2009; Longino 1990). We may not be as interested in determining the function of the institution of Pokémon trading card exchanges as that of kawaii culture more generally. However, what the function of kawaii culture is should not just be a matter of what social structures we happen to focus. We want to use functional ascriptions to make claims about the nature of social institutions that go beyond our personal or social commitments, but speak to these institutions in and of themselves.

2.6.3 The Account Should Be Actualist, Not Counterfactual

To be compelling an account of social functionalism should not *just* depend on counterfactual judgments *per se*. Of course, counterfactual reasoning is pervasive in science, social science, and beyond (Fodor 1990; Kusch 2005), and it is plausible that by saying anything about will or does happen, one implicit makes claims about what would happen under certain counterfactual circumstances (Stalnaker 1968; Lewis 1973). However, it is not compelling for functional ascriptions in the social sciences to just depend on what would happen in various counterfactual situations (such as the arrival of factors threatening the existence of a given social institution). These kinds of counterfactual judgements are often very hard to justify and assess, thus threatening to render functional ascriptions in the social sciences highly insecure. Instead, such functional ascriptions should be based on.[11]

2.6.4 The Account Should Be General

An account of social functionalism should apply to all social institutions—at least in principle. It should not just be applicable to a small subset of the latter (e.g. ones with an intentional designer), but be useable across the board. This is meant to be a major approach towards social analysis, not a minor tool to be used in a select few cases only.[12]

In the next chapter, I shall lay out and defend an account that satisfies these desiderata.

References

Ariew, A., R. Cummins, and M. Perlman, eds. 2002. *Functions: New Essays in the Philosophy of Psychology and Biology*. Oxford University Press.
Bigelow, J. C. 1998. Functionalism in Social Science. In *Routledge Encyclopedia of Philosophy*. Taylor and Francis.
Birch, J. 2017. *The Philosophy of Social Evolution*. Oxford: Oxford University Press.
Boyd, R., and P. Richerson. 2005. *The Origin and Evolution of Cultures*. Oxford: Oxford University Press.
Brandon, R. 1990. *Adaptation and Environment*. Princeton: Princeton University Press.
Campbell, C. 1982. A Dubious Distinction? An Inquiry into the Value and Use of Merton's Concepts of Manifest and Latent Function. *American Sociological Review* 47 (1): 29–44.
Cummins, R. 1975. Functional Analysis. *The Journal of Philosophy* 72 (20): 741–765.
David, P. 1986. Understanding the Economics of QWERTY: The Necessity of History. In *Economic History and the Modern Economist*, ed. W. N. Parker, 30–49. Basil Blackwell.
Den Nieuwenboer, N. A., and M. Kaptein. 2008. Spiraling Down into Corruption: A Dynamic Analysis of the Social Identity Processes That Cause Corruption in Organizations to Grow. *Journal of Business Ethics* 83 (2): 133–146.
Dipert, R. 1993. *Artifacts, Art Works, and Agency*. Philadelphia: Temple University Press.

---

[11] Again, this is consistent with the fact that the appeal to features of the social system as it actually is entails a number of counterfactuals. The point is just the latter do not ground the relevant functional ascriptions—only the former does.

[12] As will be made clear in the next chapter, this does not mean that it needs to be able to answer all questions in the social sciences. The point is just that its domain should be all social institutions, not just some of them.

Dipert, R. 1995. Some Issues in the Theory of Artifacts: Defining 'Artifact' and Related Notions. *The Monist* 78 (2): 119–135.
Doolittle, W. F. 2014. Natural Selection Through Survival Alone, and the Possibility of Gaia. *Biology and Philosophy* 29:415–423. https://doi.org/10.1007/s10539-013-9384-0.
Douglas, H. 2009. *Science, Policy, and the Value-Free Ideal*. Pittsburgh: University of Pittsburgh Press.
Durkheim, E. 1915 [1971]. *The Elementary Forms of the Religious Life* (J. W. Swain, Trans.). London: Allen & Unwin.
Elster, J. 1979. *Ulysses and the Sirens*. Cambridge: Cambridge University Press.
Fodor, J. 1990. *The Theory of Content*. Cambridge, MA: MIT Press.
Fox Keller, E., and H. E. Longino, eds. 1996. *Feminism and Science*. Oxford University Press.
Fresco, N., E. Jablonka, and S. Ginsburg. 2018. The Construction of Learned Information Through Selection Processes. In *The Routledge Handbook of Evolution and Philosophy*, ed. R. Joyce. Routledge.
Garson, J. 2012. Function, Selection, and Construction in the Brain. *Synthese* 189:451–481.
Godfrey-Smith, P. 2009. *Darwinian Populations and Natural Selection*. Oxford: Oxford University Press.
Henrich, J. 2015. *The Secret of Our Success: How Culture Is Driving Human Evolution, Domesticating Our Species, and Making Us Smarter*. Princeton, NJ: Princeton University Press.
Henrich, J., and R. McElreath. 2007. Dual-Inheritance Theory: The Evolution of Human Cultural Capacities and Cultural Evolution. In *The Oxford Handbook of Evolutionary Psychology*, ed. R. Dunbar and L. Barrett, 555–570. Oxford University Press.
Heyes, C. M. 2018. *Cognitive Gadgets: The Cultural Evolution of Thinking*. Cambridge, MA: Harvard University Press.
Hilpinen, R. 1993. Authors and Artifacts. *Proceedings of the Aristotelian Society* 93:155–178.
Hodgson, G., and T. Knudsen. 2010. *Darwin's Conjecture*. Chicago: University of Chicago Press.
Holmwood, J. 2005. Functionalism and Its Critics. In *Modern Social Theory: An Introduction*, ed. A. Harrington, 87–109. Oxford University Press.
Houkes, W., and P. Vermaas. 2004. Actions Versus Functions: A Plea for an Alternative Metaphysics of Artifacts. *The Monist* 87 (1): 52–71.
Houkes, W., and P. Vermaas. 2010. *Technical Functions: On the Use and Design of Artefacts*. Dordrecht: Springer.
James, J. E., P. G. Nelson, and J. Masel. 2023. Differential Retention of Pfam Domains Contributes to Long-term Evolutionary Trends. *Molecular Biology and Evolution* 40 (4): msad073. https://doi.org/10.1093/molbev/msad073.
Kincaid, H. 1990. Assessing Functional Explanations in the Social Sciences. *PSA: Proceedings of the Biennial Meeting of the Philosophy of Science Association* 1990:341–354. http://www.jstor.org/stable/192715.
Kincaid, H. 2009. A More Sophisticated Merton. *Philosophy of the Social Sciences* 39 (2): 266–271. https://doi.org/10.1177/0048393109333441.
Kley, R. 1994. *Hayek's Social and Political Thought*. Oxford: Oxford University Press.
Kusch, M. 2005. Fodor v. Kripke: Semantic Dispositionalism, Idealization and Ceteris Paribus Clauses. *Analysis* 65 (2): 156–163.
Lessig, L. 2013. "Institutional Corruption" Defined. *Journal of Law, Medicine & Ethics* 413:553–555.
Lewin, P. 2001. The Market Process and the Economics of QWERTY: Two Views. *The Review of Austrian Economics* 14 (1): 65–96. https://doi.org/10.1023/A:1007811722566.
Lewis, D. 1973. Causation. *The Journal of Philosophy* 70:556–567.
Longino, H. 1990. *Science as Social Knowledge*. Princeton: Princeton University Press.
McLaughlin, P. 2000. *What Functions Explain: Functional Explanation and Self-Reproducing Systems*. Cambridge: Cambridge University Press.
Merton, R. 1968. *Social Theory and Social Structure*. New York: Free Press.

References

Miller, S. 2017. *Institutional Corruption: A Study in Applied Philosophy*. Cambridge: Cambridge University Press.
Millikan, R. 1984. *Language, Thought, and Other Biological Categories*.
Millikan, R. 1990. Truth Rules, Hoverflies, and the Kripke-Wittgenstein Paradox. *The Philosophical Review* 99 (3): 323–353.
Millikan, R. 2002. *Varieties of Meaning*. Cambridge, MA: MIT Press.
Neander, K. 2006. Content for Cognitive Science. In *Teleosemantics*, ed. G. F. Macdonald and D. Papineau. Oxford University Press.
Nichols, S. 2004. *Sentimental Rules: On the Natural Foundations of Moral Judgment*. Oxford: Oxford University Press.
North, D. C. 1990. *Institutions, Institutional Change, and Economic Performance*. Cambridge: Cambridge University Press.
Novick, R. 2023. *Structure and Function*. Cambridge: Cambridge University Press.
Papineau, D. 1987. *Reality and Representation*. Oxford: Blackwell.
Papineau, D., and J. Garson. 2019. Teleosemantics, Selection and Novel Contents. *Biology and Philosophy* 34 (3)., Article 36. https://doi.org/10.1007/s10539-019-9689-8.
Parsons, T. 1951. *The Social System*. London: Routledge.
Pettit, P. 1996. Functional Explanation and Virtual Selection. *The British Journal for the Philosophy of Science* 47 (2): 291–302.
Piccinini, G., and J. Garson. 2014. Functions Must Be Performed at Appropriate Rates in Appropriate Situations. *British Journal for the Philosophy of Science* 65 (1): 1–20.
Potochnik, A. 2017. *Idealization and the Aims of Science*. Chicago: University of Chicago Press.
Potts, R., K. Vella, A. Dale, and N. Sipe. 2016. Exploring the Usefulness of Structural–Functional Approaches to Analyse Governance of Planning Systems. *Planning Theory* 15 (2): 162–189.
Price, G. R. 1995. The Nature of Selection. *Journal of Theoretical Biology* 175:389–396.
Rappaport, R. A. 1968. *Pigs for the Ancestors*. New Haven: Yale University Press.
Rappaport, R. A. 1979. *Ecology, Meaning and Religion*. Richmond: North Atlantic Books.
Rosenberg, A. 2012. *Philosophy of Social Science*. 4th ed. Boulder, CO: Westview Press.
Satz, D. 2013. Markets, Privatization and Corruption. *Social Research* 80 (4): 993–1008.
Schulz, A. 2020. *Structure, Evidence, and Heuristic: Evolutionary Biology, Economics, and the Philosophy of their Relationship*. New York: Routledge.
Searle, J. 1995. *The Social Construction of Reality*. New York: Free Press.
Sober, E. 1984. *The Nature of Selection*. Cambridge: Cambridge University Press.
Sober, E. 2000. *Philosophy of Biology*. 2nd ed. Boulder, CO: Westview Press.
Sober, E. 2014. Evolutionary Theory, Causal Completeness, and Theism: the Case of "Guided" Mutation. In *Evolutionary Biology: Conceptual, Ethical, and Religious Issues*, ed. D. Walsh and P. Thompson, 31–44. Cambridge University Press.
Spelke, E. 2022. *What Babies Know*. Oxford: Oxford University Press.
Stalnaker, R. 1968. A Theory of Conditionals. In *Studies in Logical Theory*, ed. N. Rescher, 98–112. Basil Blackwell.
Thompson, D. F. 1995. *Ethics in Congress: From Individual to Institutional Corruption*. Washington, DC: Brookings Institution.
Thompson, D. F. 2018. Theories of Institutional Corruption. *Annual Review of Political Science* 21:495–513.
Treviño, L. K., N. A. Den Nieuwenboer, and J. J. Kish-Gephart. 2014. (Un)ethical Behavior in Organizations. *Annual Review of Psychology* 65:635–660.
Turner, D. 2007. *Making Prehistory: Historical Science and the Scientific Realism Debate*. Cambridge: Cambridge University Press.
Turner, S. 2009a. Many Approaches, But Few Arrivals: Merton and the Columbia Model of Theory Construction. *Philosophy of the Social Sciences* 39:174–211.
Turner, S. 2009b. Shrinking Merton. *Philosophy of the Social Sciences* 39 (3): 481–489.

Vermaas, P., and W. Houkes. 2003. Ascribing Functions to Technical Artefacts: A Challenge to Etiological Accounts of Functions. *The British Journal for the Philosophy of Science* 54 (2): 261–289.

Vincent, J. 2015. Functionalism in Anthropology. In *International Encyclopedia of the Social & Behavioral Sciences*, ed. J. D. Wright, 2nd ed., 532–535. Elsevier. https://doi.org/10.1016/B978-0-08-097086-8.12077-X.

Vrba, E. 1984. What is Species Selection? *Systematic Zoology* 33:318–328.

Whiten, A., and R. W. Byrne, eds. 1997. *Machiavellian Intelligence II: Extensions and Evaluations*. Cambridge University Press.

Wilson, D. S., and J. M. Gowdy. 2013. Evolution as a General Theoretical Framework for Economics and Public Policy. *Journal of Economic Behavior & Organization* 90S:S3–S10.

Open Access This chapter is licensed under the terms of the Creative Commons Attribution 4.0 International License (http://creativecommons.org/licenses/by/4.0/), which permits use, sharing, adaptation, distribution and reproduction in any medium or format, as long as you give appropriate credit to the original author(s) and the source, provide a link to the Creative Commons license and indicate if changes were made.

The images or other third party material in this chapter are included in the chapter's Creative Commons license, unless indicated otherwise in a credit line to the material. If material is not included in the chapter's Creative Commons license and your intended use is not permitted by statutory regulation or exceeds the permitted use, you will need to obtain permission directly from the copyright holder.

Chapter 3
Presentist Social Functionalism—The Foundations

Abstract This chapter develops a new account of social functionalism: presentist social functionalism. In particular, the chapter argues that social functionalism should be based on the *actual* bio-cultural selective or sorting pressures on the institution *as it is now*. The chapter shows how this idea can be deepened and underwritten with recent insights from evolutionary biology and the philosophy of biology. It also considers a number of the key objections to this kind of view that have been or could be proposed, and shows why they fail to be compelling.

3.1 Introduction

As I show in this chapter, it is possible to provide a plausible account of functional ascription in the social sciences that satisfies the desiderata just laid out. This account grounds the function of social institutions in those of their features that, in the current cultural environment, increase the chances of these institutions to survive or reproduce. Such a presentist account is shown to be able to make clear that the function of many social institutions may be more complex than hitherto assumed, and that existing discussions of these functions have often focused on the wrong kinds of empirical data.

As pointed out in the introduction, the best way to see how this account works is by putting it into practice. This will therefore be the target of the next three chapters. However, in order to provide a compelling foundation for these applications, it is necessary to begin by (i) spelling out the basic structure of the account, thus making clear how it meets the desiderata of the previous chapter, and (ii) responding to some possible worries concerning the structure of the account. The goal in what follows is thus to display the promise of account to improve on the existing accounts and make its fundamental machinery clear. The rest of the book will then deliver on this promise and show how this machinery can be put into practice in specific cases.

The chapter is structured as follows. In Sect. 3.2, I develop the novel form of functionalism that is the core of the present book, and show how this account meets the desiderata of the previous chapter. In Sect. 3.3, I respond to some worries for the account. I summarize the discussion in Sect. 3.4.

3.2 Presentist Social Functionalism

The account of social functionalism to be developed here uses some of the same core insights at the heart of the accounts of Bigelow (1998); Kincaid (1990); and Pettit (1996)—and indeed that of Merton (1968)—but spells these insights out in a different way. In particular, the present account shares with structural functionalism and the account of Pettit (1996) the idea that social functionalism should not be grounded in the history of the relevant social institutions (whether that history is design-based or selection-based in one sense or another). However, unlike the account of Pettit (1996), the present account does not look towards how a social institution would react to various counterfactual scenarios to ground functional ascription to it, but towards the *actual* bio-cultural selective or sorting pressures on the institution *as it is now*.[1] In this regard, the present account has more in common with that of Bigelow (1998) (and see also Kincaid 1990)—however, the way this idea is spelled out here is very different from the (rather brief) hints in Bigelow (1998).

Specifically, the key idea of the account to be defended here is that a social institution N has function F if it is *now* selected or sorted for F.[2] That is, the question at the heart of the account is: does feature F of social institution N make it more likely that N *will* survive or reproduce in the current socio-cultural environment? If it does, F is (part of) the function of N; if not, it is not. The present account thus makes the function of N dependent on those features of N—if any—that increase the expected survival or reproductive success of N in its current environment. More precisely:

Presentist Social Functionalism: Feature F of social institution N is (part of) the function of N if F makes it more likely that N will survive or reproduce in the current socio-cultural environment.

To understand this better, a number of further points about this account need to be noted.

First, the present account explicitly and intentionally groups together genuine selection—i.e. the heritable differential reproduction of social institutions—and mere sorting—i.e. the differential growth or persistence of social institutions. There are two reasons for this.

On the one hand, as noted in the previous chapter, a persistence-based view of selection may in fact be in line with evolutionary biological theorizing: in particular,

---

[1] Some passages of Pettit (1996) seem to emphasize the current adaptive pressures on a given social institution, too. However, it is clear (e.g. from looking at its title) that the main focus of the latter account is on virtual selection, as sketched above.

[2] In a slightly different form, the focus on which traits are *adaptive*—rather than *adaptations*—to ground function has been put to use in other contexts, too (see e.g. Nanay 2014); however, the present form of this idea is unique and novel.

there may be good evolutionary biological reasons for seeing mere persistence and differential reproduction as merely different instantiations of the same kind of phenomenon (Doolittle 2014; Price 1995; Fresco et al. 2018).

On the other hand, even to the extent that this is denied—that is, even to the extent that only differential reproduction of entities in a population is seen as truly a case of natural selection—for present purposes, this is not problematic. The goal of the present account is to ground the function of social institutions. As also noted in the previous chapter, while an appeal to the grounding of functions in biology is useful as a starting point, there is no requirement that these two groundings need to be the same. Social science has different (if at least sometimes related) explanatory goals than the biological sciences, and so some divergence in their core notions need not be greatly problematic. In the present context, the divergence results from the fact that it is not generally plausible to see social institutions as reproducing. However, the differential *growth or survivorship* of different kinds of social institutions *is* generally plausible. So, when it comes to social functionalism, the focus *should* be on the latter kind of process (though, as noted in the previous chapter, the former need not be ruled out a priori either): this is what allows us to see certain institutions as being more stable parts of a given society. Hence, the fact that social functions can be grounded in either the sorting or is made explicit on the present account.

Second, in virtue of the non-historical nature of the present account, stating that social institution N has function F should not and cannot be taken as an explanation of why the institution is in existence now. Rather, N's having F is forward looking: it expresses why we should become more confident that N will continue to exist in the future. However, this is again in line with what a compelling version of social functionalism *should* look like (Bigelow 1998). As noted in the previous two chapters, one of the core aims of social functionalism is to explain why some social institutions are more stable parts of a given society than others. This is precisely what the present account can do. By showing that certain social institutions have features (their function) that make their survival or reproduction more likely, we are put in a position to hone in on the institutions that are more likely to be around in the future, too.[3]

Third, the role of counterfactuals is quite different on the present account as compared to that of Pettit (1996). On the present account, the only counterfactual that matters is whether the expected reproductive or persistence-focused success of social institution N would decrease if it did not have F.[4] We do not need to consider whether N with F would continue to exist in all nearby possible worlds. We thus do not need to imagine all sorts of scenarios that may threaten N's existence. The question is just whether feature F contributes to N's expected survival or reproductive success as it is now. Typically—though, as will also be made clearer in the next

---

[3] This is thus something that the present account shares with that of Pettit (1996).
[4] Indeed, precisely the latter is at heart of Nanay's (2014) non-etiological account of biological functioning.

section, not necessarily—this can be much more easily assessed, both theoretically (e.g. using background knowledge about the causal structure of the relevant economic system) and empirically (e.g. using carefully designed randomized controlled trials). In this way, we can sidestep the key problems that befall the account of Pettit (1996).

Fourth, the present account also sidesteps the problems of the historically-focused versions of social functionalism. Since the present account is not historical, it does not matter that many social institutions do not emerge after being "tried out," in different versions, in one or several cultures. Similarly, it does not matter that social institutions are often relatively young and thus lack the kind of history that would make a purely persistence-based view of selection (as in Doolittle 2014) plausible to ground their function, or that their history would make such functional ascriptions ambiguous and unclear. Here, the past selective or other history of the institution does not matter at all to its functional ascription. In this way, the present account can take the best from historical selectionist accounts—namely their focus on actuality, rather than counterfactuals—but does not face their worries. In particular, it does not matter for the present account if there is not institutional variation in the current social environment: the present account asks just if feature F increases the expected survival or reproductive success of institution N. The comparison here is not with other existing social institutions or different time-slices of the current institution, but with a (possibly) counterfactual version of the *present* social institution that lacks F. This appeal to the present thus avoids the ambiguity problems plaguing some of the historical version of social functionalism.

For the same reason, the present account is also entirely consistent with the fact that the evolution of many social institutions is heavily dependent on chance. The core of the present account is just that N's having F *increases* the expected survival or reproductive success of N. The present account does not claim that N's having F *fully determines* the survival or reproductive success of N. It is not a claim about the actual evolution of N—just about one of the factors that influences this evolution in the current bio-cultural environment.

The fifth point worth noting here is that the present account can still allow for malfunction—a major requirement of any plausible account of what grounds functional ascription (Millikan 1984, 1990; Neander 2006; Fodor 1990). It is not like anything that N does is part of its function. Rather, only those features that contribute to its expected reproductive or survival success are part of this function. So, to see a social institution N (ritualistic pig slaughter, say) as having function G (to entertain children, say) might turn out to just be *wrong*: while G may indeed be a feature of N (children may be entertained by the ritual), unless G increases N's expected reproductive or survival success, it is not its function. In this way, the present account can be seen to be a defensible way to ground functions.[5]

---

[5] One could also consider a different sense of "malfunction," according to which we identify features a given social institution lacks, but which are thus that, if it were to have them, would increase its expected reproductive or survival success. However, it is not so clear to what extent appealing

Sixth and finally, according to the present account, a social institution N could have more than one function.[6] So, it may be that there is a set of features F_1 to F_n that, individually, increase the expected reproductive or survival success of N. In that case, all of F_1 to F_n would be the function of N. Also, it may be that some feature F increases the expected reproductive or survival success of N but only in the right circumstances—e.g. if there are not too many other tokens of N with F in the population (as in situations of frequency-dependent selection in biology: see e.g. Gillespie 1998; Futuyma 2009). These points will become important again momentarily. For present purposes, though, it is sufficient to note that the present version of social functionalism allows for—and indeed *invites*—complex, multi-faceted and dynamic functional ascription in the social sciences. However, importantly, it does so in a non-ambiguous way: it is not that it is not *clear* whether F_1 to F_2 should be taken to be the function of N, as was true on persistence-based versions of historical functionalism. That is, the issue is not that at some point in the past, F_1 increased the survival chances of N, and at some point in the past, F_2 did, thus making it ambiguous which of these is the current function of N. Rather, the claim is that *both* F_1 and F_2 increase N's expected reproduction or survival success *now*. Despite its pluralism, therefore, the present form of social functionalism is non-ambiguous.

All in all: according to the present account, a social institution N has function F if F increases the expected reproductive or survival success of N in the present cultural environment. Functional ascription in the social sciences is thus shown to be ahistorical and to just be based on comparing the expected reproductive or survival success, in the current bio-cultural environment, of N with F to that of N without F. In this way, the present account is well placed to satisfy the four desiderata revealed by the discussion of the existing accounts. Of course, as noted before, the best way to really bring out the power of the account is by applying to some concrete questions in the social sciences. Before doing this, though, it is useful to note that all the pieces are in place here for a successful treatment of social functionalism.

3.2.1 The Account Should Be Focused on the Present

This is a defining feature of the present account. As just noted, this allows it to make sense of the fact that social institutions tend to be too recently created and lack the kinds of features—variation, heritability, and reproduction—to make appeal to historical forms of selection compelling. Apart from the fact that appealing to features of the present avoids the difficulties of justifying historical claims (Turner 2007) (as noted in the previous chapter), there is another benefit of this present-focus that is

---

to this kind of counterfactual sense of "malfunction" is useful in the social sciences; accordingly, it will not be central in what follows.

[6] This is a point that the present account shares with that of Merton (1968).

worth making explicit here. This concerns the fact the present account can avoid "swampman-"type worries (Millikan 1996; Neander 1996): a social institution N may be structured identically and be identically situated in its social system compared to another social institution N', but lack the requisite history to have a function. On historical versions of functionalism, only N' would have function—even though N is identical to it. The present account, though, is immune to this kind of worry: functional ascription independent of historical information, and we thus do not need to worry about the history of the institution—or what we know of it.

This matters, for while "swampman"-type worries may not be particularly pressing in biology (Neander 1996), they are not so easy to brush aside in the social realm. As also noted in the last chapter, social institutions can arise quickly, and it is entirely possible than an institution N_1 (making music with string instruments, say) appears in a society S_1 that has much of the same features F as a much older social institution N_2 in society S_2, which has furthermore been (culturally) selected for having F (producing continuously manipulable notes at a volume conducive to smaller gatherings). In those cases, it can be theoretically and empirically problematic (and not merely "counterintuitive") if N_2 is said to have function F, but N_1 is thought to be functionless. From a social scientific standpoint, there can be good reason to treat these two cases similarly—especially if F makes both N_1 and N_2 more likely to be able to reproduce or survive, and thus be a continued element of their respective societies.

3.2.2 The Account Should Be Non-arbitrary

The selection and sorting pressures on a social institution N—i.e. the set of features F of N that increase the expected reproductive or survival success of N—are not observer-dependent in an arbitrary way. Of course, the grain of detail with which we describe N or F is dependent on our theoretical interests, moral, political, and aesthetic interests (a point to which I return in the next section). We can describe the function of ritualist pig slaughter as food security maintenance, agricultural yield maximization, or soil preparation. Similarly, which social institutions we choose to analyze—whether it is Pokémon trading card exchange or kawaii culture—depends on our personal, social, and theoretical commitments. However, the functional ascriptions we make are constrained by the nature of the social institutions in question, and cannot be freely chosen. They are thus non-arbitrary in the desired way.[7]

---

[7] This relates to the point made in Chap. 2 that structural functionalism is observer-dependent in exactly this problematic manner. Here, as there, the point is not that we need to avoid all kinds of observer-dependence (most basically, functional *ascription* has to be observer-dependent in the sense that it requires someone to actually do the ascribing). Rather, the point is that the present account makes functional ascription more than a matter of choice: it needs to actually respect the socio-cultural facts.

3.2.3 The Account Should Be Actualist, Not Counterfactual

By basing functional ascription on those features F of a social institution N that increase its expected reproductive or survival success of N, presentist social functionalism is inherently actualist. Of course, as also noted earlier, there are some implicit counterfactual commitments in the account: to say that F increases the expected reproductive or survival success of N, we are saying that if N had lacked F, its expected reproductive or survival success would have been lower. However, this does not take away from the fact that the present account is strongly focused on the actual world: we are not considering what would happen to N *just* in other possible worlds—we are just interested in what N does here, in the actual world. It is just that the latter has to be underwritten with some counterfactuals. However, it is the claim about the actual world that matters, not the features of N in various worlds that are independent from the actual world. I return to this point in the next section.

3.2.4 The Account Should Be General

There are no content-restrictions built into presentist social functionalism. It can be applied to all social institutions—even ones in the past. (Indeed, as will be made clearer in Chap. 6, the account can be extended to functional ascriptions to artifacts.) The point is just that functional ascription is relativized to a given time: the function of N at t is whatever set of features F of N increases the expected reproductive or survival success of N at t.

In all therefore: presentist social functionalism—the view that the function(s) of social institution N consist(s) in the set of features F of N that increase its expected reproductive or survival success—is well-placed to improve on the other accounts of social functionalism in the literature: it is non-historical, non-arbitrary, actual, and general.

While the remaining chapters of the book will illustrate this in more depth, it is at this point useful to show how the present account improves our understanding of a classic problem issue in the social sciences (especially anthropology): the function (or lack thereof) of Chinese foot-binding. The literature on foot-binding practices is very large and variegated (see e.g. Brown 2016; Ko 2007; Blake 1994), but for present purposes, it is sufficient to note that there is much controversy over whether foot-binding has a function at all, and, if it does, what its function actually is. Here, the presentist social functionalist treatment is helpful, as it can correct, clarify, and advance the existing discussion.

So, some authors have suggested that foot-binding is a "culturally arbitrary fashion" (Shepherd 2018; see also Ko 1997). Here, what the presentist account makes clear is that just because something is a "fashion" does not mean that it is functionless: wearing white on one's wedding day (say) could spread due to its signaling commitment to traditional social values—and thus have this function—even though it is otherwise "culturally arbitrary." Similarly, presentist social functionalism shows that the function of foot-binding can be complex and go beyond single-factor explanations—whether based on the pressures exerted by a mating market (Mackie 1996) or other social, ethnographic, economic pressures (see Shepherd 2018, for an overview). This is helpful not only because multi-factor explanations of the emergence and distribution of foot-binding are increasingly recognized as important (Tran 2020), but also because it makes it possible to see the *function* foot-binding as a combination of all of these different factors, and as something that can differ in different geographic regions or historic periods. This will become important again in the next chapter.[8]

3.3 Objections and Responses

So that presentist social functionalism can be seen to be truly credible contender in the discussion of social functionalism, it is necessary to respond to a set of objections that may come to mind. These objections fall into two categories. In the first group are objections that target the *basic, defining features* of the account. These objections tend to derive from similar worries for related views in other contexts, such as evolutionary biology or cognitive science. In the second group of objections are those that target *specific* features of the account, as it is spelled out here. These objections are thus unique to this context; while they might still have mirror images in other contexts (such as evolutionary biology and cognitive science), this is not their origin. While there are of course many possible objections of either kind that could be raised here, the following discussion will consider two of the most central objections falling under each of the above two classes. This ensures that this discussion remains manageable without sacrificing too much in the way of depth or breadth.

Before beginning this discussion, though, it is important to emphasize that aspects of the objections and replies will also be picked up again in the ensuing chapters. As noted above, the goal in this chapter is just to show that presentist social functionalism can get off the ground in the first place. In this sense, the discussion of the objection and replies in this chapter also helps situate and the discussion in the chapters to follow.

---

[8] In the same way, presentist social functionalism can build on Hayek's insight that many important social institutions are "spontaneous orders" (i.e. not intentionally designed) (Hayek 1967, 1973; Vaughn 2013; Garrouste 1994). In particular, the account defended here shows that social institutions can spread as they foster coordination and cooperation, independently of whether participants recognize that they do so. See also Chaps. 4 and 6 for more in this.

3.3 Objections and Responses

3.3.1 General Objections

The first foundational objection to consider here concerns the related claims that (a) presentist social functionalism has counterintuitive consequences and (b) presentist social functionalism answers the wrong sorts of questions. In particular, given the presentist focus of the account, it would seem that the account implies that the function of various social institutions is or could be very different from what it is commonly taken to be. For example, consider the police force again. If, in some particular social context, it turns out that the police force of a given state is persisting because it is being used to suppress dissent, then, on the present account, that would now be seen to be its function. This may seem odd, since we may also want to say that the function of the police *really* is to uphold law and order (say)—especially if that is what it was founded or culturally selected for in that past. It is just that (we may be tempted to go on to say), in the (supposed) case in question, the police force is doing something that is not in line with what it is for. Indeed, we may want to use the fact that the police force is failing to act in ways that match its historical "function" (i.e. that it suppresses dissent rather than upholding the law) as grounds for criticizing these actions. However, it may appear that this is not something that we can do with presentist social functionalism in the background: if suppressing dissent is what increases the chances of survival and reproduction of the police force, then that is its function. (As should be clear, this worry is related to the worries discussed in the context of structural functionalism in the previous chapter.)

Relatedly, as far as claim (b) above is concerned, with presentist social functionalism, we are unable to *explain* why a social institution operates in the ways it does. Since this account is not historical, it cannot answer the question of the origin of the institution. However, explaining why a social institution does what it does—e.g. how a given police force came to suppress dissent—would seem to be something we want to do.

In response, three points need to be noted. First, the counterintuitive implications of the account are in fact to be embraced. The account is revisionist in structure, and is not trying to capture "our" pre-theoretic intuitions about the functions of social institutions. This would anyways be difficult, since it is far from clear what these intuitions are, nor is it clear who the "we" is that is meant to share them (All humans? All citizens of a given state? All political scientists?). Rather, the presumption behind presentist social functionalism is that the *function* of a social institution is a technical, social scientific concept that we can choose to structure in ways conducive to our social scientific investigations. Indeed, it is precisely this which has led to the formulation of the four desiderata from the previous chapter: we want an account of social functions that is presentist, actualist, non-arbitrary, and general for the reasons laid out there. We are not strictly beholden to whatever we or someone else may have thought about the function of a given social institution. Of course, the latter is a useful data point to consider, but it should not be seen as a non-negotiable constraint. In other words, the goal of the project is not a summarizing description of the foundations of our social functionalist intuitions, but a theoretical account to be used in the social sciences and philosophy. The former may of course also be an interesting project; it is just not the one at stake here.

Second, this does not mean that it is not possible to criticize the way a social institution operates. It is just that the nature of this criticism changes. On the one hand, it is still possible to criticize a social institution for not fulfilling its function: we are just spelling out this function in a different way here. This is an important point to which the next three chapters return (especially Chap. 5). For now, though, it is sufficient to note that presentist social functionalism does not mean that social institutions cannot be criticized. It is true that we cannot criticize the above police force for not fulfilling its function by suppressing dissent. However, we can criticize the police force for doing other things—taking bribes (assuming to does so), say.

On the other hand and very importantly, instead of noting that a social institution does not fulfill its function, we can also criticize a social institution for having the function it does. So, if a police force persists because it is being used to suppress dissent, then we can say that we are now living in a society where the police force has a function it should not have.[9]

This can be illustrated well by the discussion surrounding human rights. Instead of starting with a fixed conception of what humans rights are and of who has them, we could say: if a given set of laws surrounding human rights is problematic—e.g. because it affords certain members of the species homo sapiens (African Americans, say, or females) fewer protections than others—then that is a reason for criticizing these laws. This would not be because the laws do not match existing human rights, but because the laws set up the wrong kinds of "human rights" (Darby 2003, 2004; Martin 1993). To be sure, the latter treatment of human rights is very controversial (see e.g. Feinberg 1992; Dworkin 1977; Nickel 1987), but there is no need to engage with this discussion here. Whatever may be the case when it comes to human rights—an issue that heavily depends on the nature of legal and moral systems— what matters for present purposes is just that these discussions illustrate the fact that a revisionist account of social functions is not *required* to be evaluatively quietist. A parallel treatment may be plausible when it comes to human rights, but it is certainly compelling when it comes to the much less morally and legally loaded question of the function of social institutions. We can criticize social arrangements for making it the case that certain institutions have certain functions just as much as criticizing them for making institutions act in ways that fail to match their function. Both forms

---

[9]A related worry is that it may appear not so clear to what extent a social institution with the function of suppressing dissent is really a *police force*. However, as made clearer in Chap. 1 in the context of the discussion of the teleological account of social institutions and below in the discussion of the individuation of social institutions, the present account does not build functions into the individuation of social institution. There may be all sorts of reasons why we want to individuate an institution as a "police force" even if it has the function to suppress dissent (including its historical genesis, charter, and how it is commonly labeled in the relevant society). Similarly and for the same kinds of reasons, why we may see two (token) institutions, in different cultures perhaps, as instances of the same type of institution even if they have different features—including different functional features.

3.3 Objections and Responses

of criticisms are legitimate—and both can be accounted for on the presentist social functionalist viewpoint.

Third, it is true that presentist social functionalist theorizing cannot be used to explain where a given institution is coming from. However, this is not greatly problematic. It is of course still true that we can inquire into the historical presuppositions and development of a given social institution. It is just that this is not built into the functional analysis of the institution per se, but treated as a separate inquiry. Importantly, it is far from clear that this division is at all problematic. As also noted by Pettit (1996), functionalist analysis is typically used to identify the stable social institutions—the ones that are likely to persist into the future. It is not at all obvious that what we also want to use *social functionalist analysis* to identify why a social institution came to be structured the way it is. Of course, we may want to know the answer to the latter question, but, at the very least, it is far from obvious that an account of social functionalism needs to provide it. On the contrary, not using social functionalist analysis to answer the latter question is not a major departure from the major forms of functionalist theorizing in the social sciences.

The second general objection to consider here concerns the fact that, on presentist social functionalism, determining the function of a social institution is very difficult. After all, it often seems very unclear which features of a social institution increase its expected reproductive or persistence success—and even where we have evidence about this set of features, it would seem to be subject to frequent change. A small change in the social environment can make it the case that what was once reproductive or persistence success-enhancing is no longer so, and vice versa. For example, consider a state where the police force is used to suppress dissent as well as to uphold law and order. What is the function of the police force? How can we tell which of these features increases its expected chances of reproduction or persistence? Moreover, it would seem that a small change—the promotion of a new police chief, say—can change this function quickly. In short: the worry is that, in virtue of its fundamental structure, presentist social functionalism is hard to apply to actual cases: due to the difficulties with ascertaining the often changing features of a social institution that increase its expected chances of reproduction or persistence in the present cultural climate, functional ascription appears to become difficult or impossible.

In response, there are again three things to note. First, since the proof of the pudding is in the eating, the best way to show the applicability of presentist social functionalism is by actually applying it. This is therefore one of the main goals of the next three chapters. As will become clear there, it turns out that the applicability of presentist social functionalism is actually quite high, and far from an impossibility.

The second point to note in response to this objection is that this concern is not specific to presentist social functionalism, but in fact affects all versions of social functionalism. As noted, earlier, much the same questions (concerning the features that increase an institution's expected reproductive or persistence success) need to be answered for the historical accounts—only with the added difficulty that obtaining evidence of the past is often tricky. In the present, we can design experiments (at least sometimes) and generally access social institutions in ways we cannot when it comes to the past. For structuralist functionalism, the problem is slightly different, but still related: we need to determine the contribution an institution makes to a given social arrangement. This also need not be obvious (and, as noted earlier, it faces the arbitrariness worry that we can choose which social arrangement to focus on). Finally—and as also noted earlier—for Pettit's counterfactual account, determining which features of an institution would make it the case that the institution would persist in cases where its existence were threatened is very difficult, too: assessing such counterfactuals is far from easy. Putting this the other way around: far from it being a problem for presentist social functionalism, the applicability of the approach is, if anything, a benefit of the latter when compared to the alternative approaches.

That said, the objection is correct in pointing out that determining the features that increase and institution's expected chances of reproduction or persistence need not be easy. However, this is a challenge to overcome, not a deep objection to the account. In fact, the same can be said for most other sciences: determining the nature of dark matter or the genetic bases of complex traits is not easy either. This, though, does not mean it is impossible either. The same is true here: while it may be a challenge to determine the features that increase and institution's expected chances of reproduction or persistence, doing so need not be impossible. For example, while it may be hard to determine whether a specific police force's chances of reproduction or persistence are increased by its being used to suppress dissent or the fact that it upholds law and order, there may well be ways to assess this: we may be able to consider whether there are other institutions that uphold law and order (the armed forces, say), or whether the suppression of dissent in fact increases the pressure on the police force to be abandoned.[10] The following chapters further underwrite this point; but for now, it is just important to note that the application of presentist social functionalism, while non-trivial, is not impossibly difficult either.

The third point to note in response to this objection is that it is not a vice of presentist social functionalism that the function of an institution can change quickly—it is a virtue. Given that the aim of functionalist analysis is the establishment of which social institutions are stable features of a society, and which merely transitory, being attuned to the dynamic interactions between an institution and its wider social environment is crucial: it helps determine which of these two sides particular institution falls on. Again, this is something that future chapters will come back to and make more precise. For now, though, it is just important to note that it is not problematic that presentist social functionalism sees the function of social institutions as quickly changing, but something that can just be accepted as a positive feature of the account.

All in all, therefore: there are no general reasons for thinking that presentist social functionalism cannot get off the ground as a compelling approach towards

---

[10] Indeed, there is much work in applied economics that is testament to the possibility of testing claims like these: see e.g. Acemoglu et al. (2001).

3.3 Objections and Responses   47

functionalist analysis in the social sciences. It is a revisionist proposal that does not try to capture "our" intuitions about social functions (whatever these may be) and which does not try to explain the origin of a given social institution (though it is consistent with such explanations). It also is no worse off in terms of its ability to ascertain the function of a given social institution (vis-à-vis rival accounts) and can just accept the fact that the function of a social institution can change quickly. With this in mind, consider some more specific objections to the details of presentist social functionalism, as it was spelled out in the previous section.

3.3.2 Specific Objections

The first of the specific objections returns to the fact that one of the key advantages of presentist social functionalism, when compared to Pettit's counterfactual alternative, is that it is based on what is *actually* the case. However, as also noted earlier, it does not seem that the former approach can really avoid all counterfactuals: it seems we still need to assess what would have happened to the relevant institution if it did not have features F (which are meant to be its function). In what sense, therefore, does presentist social functionalism really improve over the counterfactual alternative? More generally: exactly what is the role of counterfactuals in this proposal?

In response, two points can be noted. In one sense, the objection is exactly right: presentist social functionalism does not forgo all reliance on counterfactuals. However—as also noted earlier—this is not surprising, since this would not be possible. Many claims about the actual world imply claims about counterfactual worlds (Lewis 1973; Stalnaker 1968; Kusch 2005; Fodor 1990): for example, causal claims and conditional claims—indeed, many relational claims. This includes claims about the impact a set of features has on the expected reproductive or survival success of a given social institution. The issue is not whether all counterfactuals can be avoided, but what the role is of the counterfactuals in the account. On Pettit's approach, the function of a social institution is made dependent on the persistence of the institution in counterfactual—non-actual circumstances. On presentist social functionalism, the function of a social institution is made dependent on the persistence or survival of the institution in the actual world—it is just that this is being assessed by comparing it to the actual world.

This leads to the second key point to note here. While presentist social functionalism thus cannot avoid appealing to counterfactuals altogether, the kinds of counterfactuals it appeals to are different. In particular, they are different in two dimensions. On the one hand, they are different in content. On Pettit's counterfactually-driven functionalism, *all sorts* of counterfactual scenarios would need to be considered. Would gold clubs persist in worlds with urban hiking clubs? What about in worlds with frolf (frisbee golf) clubs? What about in worlds with virtual reality metaverse golf clubs? By contrast, on presentist social functionalism, the only situation that needs to be considered is what the expected reproductive or survival success of the institution is in the nearest world where it does not have feature set F. There is thus less of a range of scenarios to consider here. This makes presentist social functionalism narrower in its counterfactual content.[11]

On the other hand, the counterfactuals that underlie presentist social functionalism are easier to assess than those underlying Pettit's alternative proposal. As noted in the previous chapter, it is often far from clear how a given social institution would respond to a specific counterfactual scenario. However, it is at least sometimes easy to determine whether the expected reproductive or survival success of a given institution would decrease without feature set F. We can compare the institution to relevantly similar institutions (including past versions of the institution in question) that lacked the feature. We may also be able to do a controlled experiment—or rely on a natural experiment—to assess the effect of removing feature set F from the institution. These are all familiar scientific inferences, and as such do not pose major methodological challenges (Sober 2008). This is especially so, since we do not need a fully precise estimate of the exact the expected reproductive or survival success of the institution with and without F; all we need is a comparative assessment of whether it is higher with or without F. This is easier to do than to assess the survival of golf clubs in a whole host of extraordinary counterfactual scenarios. Of course, it is true that assessing the expected reproductive or survival success of the institution with and without F need not *always* be straightforward. As noted earlier, there is no guarantee that this will not, at least at times, be a challenge that needs to be overcome. The point is just that, all in all, the counterfactuals are likely to be easier to assess on presentist social functionalism when compared to the fully counterfactual approach.

The final objection to be considered here picks up another point that was mentioned earlier: the individuation of institutions and features. Many things can impact the likelihood with which a social institution survives or reproduces, including the presence of other social institutions and various external features of the bio-social environment.[12] However, these do not necessarily become part of the function of a given social institution. Only if they are features *of* the institution could they be part of its function.[13] In other words, it is built into the statement of presentist social functionalism that there is an existing universe of discourse that includes the relevant social institutions and their features. What presentist social functionalism allows us to determine is which (if any) of the *given* features of a *given* social institution is its function. This may be thought to raise two concerns, though: first, where

---

[11] One way to spell this out is in terms of the *number* of counterfactuals that need to be considered on the two proposals. However, this would then require appeal to a specific modal language. This is why the text formulated in terms of the content of the counterfactuals, rather than the number of counterfactuals itself.

[12] For example, the appearance of the institution of fantasy football leagues can make the institution of the National Football League more likely to spread and persist.

[13] This point also extends diachronically: it needs to be determined when a social institution remains the same social institution and when it becomes a new one. If institution N has feature F at time t1 and a different feature G at time t2, is it still the same institution or a new one (e.g. if a company that solely produced consumer technology at t1 also starts to provide consumer lending services at t2, does it become a bank)?

3.3 Objections and Responses

are these individuation schemata coming from? Is it so clear that it is more useful to theorize about the function of "golf clubs" than about that of "sports clubs" or of "private golf clubs with membership fees that are greater than $100,000 per year?" Similarly, how do we know whether the features we should consider include "providing business networking opportunities," "spending time with like-minded people from a similar cultural background," or "learning about industry gossip?" Second and relatedly, it may be thought that it is one of the main jobs of social functionalism to help us determine what institutions to focus on in our social analysis, and what features of these institutions are particularly important. If—as seems to be the case—this is being presupposed by presentist social functionalism, this would seem to be a problem for the approach.

In response to this worry, there are three points that need to be noted. First, while the objection is correct in noting that presentist social functionalism depends on an individuation schema for social institutions and their features, it is not the case that it is the only account for which this is true. On the contrary: similar points hold for all forms of social functionalism (and indeed biological and social theorizing quite generally: see e.g. Baum 2013; Bertrand 2013; Laland et al. 2005; Odling-Smee et al. 2003; Dawkins 1982, 2004; Griffiths 2005; Griffiths and Gray 1994). In this regard, presentist social functionalism is no worse off than the other approaches mentioned in the previous chapter: it needs to be combined with some reasonable approach towards individuating society into different institutions with various features.

Second, the need for an individuation schema for social institutions and their features can be separately addressed from the question of the function and stability of social institutions. In fact, this is a very important point that returns to the objection concerning the explanation of why social institutions have the functions that they do. Social functionalism should not be seen as a panacea and thus not expected to be able to address all questions in the social sciences. The fact that presentist social functionalism needs to be combined with an individuation schema for social institutions and their features is not a *problem* for the account—it just means that this is one approach among others that can help us answer some of the questions we are looking to address. There is a tendency for functionalist approaches towards the social sciences to be seen as grand theories of everything that can "solve" the social sciences once and for all. This, though, is a tendency that ought to be resisted. Presentist social functionalism is designed to answer a specific set of questions: namely what the function of social institutions, so that we can determine which institutions are likely to be persistent elements of society in the future as well. The fact it is unable to answer every other question we might have—including the questions of why there are institutions of type X in society, why these institutions have features X, Y, and Z, etc.—is not problematic. It just means that presentist social functionalism needs to take its place among other theoretical approaches towards the social sciences.

Third, that said, this does not mean that presentist social functionalism is completely irrelevant for the determination of the right individuation schema for social institutions and their features. Rather, in line with work in non-foundationalist (social) sciences in general, presentist social functionalism can be used to help *bootstrap* such an individuation schema. We may start out with only a weakly supported such individuation schema, about which we are not sure. However, depending on the success of functionalist analysis with this schema in the background we may become surer of the latter. This is similar to what is happening in other sciences: we may hypothesize that a disease is transmitted through airborne droplets, but not be *sure* that this is the case. We can then model the spread of the disease based on this assumption. If these models then turn out to be predictively or explanatorily successful, this provides us with a reason to think that the assumption they are built on—concerning the mechanisms underlying the spread of the disease—is reasonable (Levins 1966; Fuller and Schulz 2021). Of course, there are many details to this sort of inference (Orzack and Sober 1993; Weisberg 2006; Odenbaugh 2006; Odenbaugh and Alexandrova 2011), but for now, it is sufficient to note that, exactly how it is to be spelled out, it is a familiar feature of science.

The same is true here. We may hypothesize that a given culture contains an institution of *private property* with a specific set of features (regulated rights of access and use of resources; right to transfer these rights to others); based on this, we can then ask what the function of this institution is (maybe to regulate access and use of resources). If the latter turns out to be a compelling analysis—maybe because we correctly predict that the institution of private property will persist in the future—then this might provide us with a reason for thinking our initial individuation schema was reasonable. Now, none of this is to say that providing such an individuation schema is not a separate project from (presentist) social functionalist analysis. The point is just that separate projects can still inform each other—and this is likely to be the case here also. In this way, it is not only the case that the objection here may not be much of an objection at all (even if were to grant that social functionalist analysis should help us provide an individuation schema for social institutions and their features.

3.4 Upshot

This chapter presented and defended a novel account of social functionalism: presentist social functionalism. According to this account, the functions of a social institution are those of its features that increase the probability that the social institution will survive or reproduce. This account of social functionalism is shown to improve on the major alternatives in the recent philosophical literature: structuralist functionalism, historical selection-based accounts, and the counterfactually-based account of Pettit (1996). In particular, by being grounded in the actual circumstances prevailing in society, the account defended in this chapter satisfies all the desiderata laid out in the previous chapter: it is non-historical, non-arbitrary, and non-counterfactual.

The account is also shown to be able to deal with the key objections that can and have been raised against this kind of view. In particular: (i) the account is revisionary and thus not answerable to our intuitions about specific cases (whatever these are), nor is it designed to explain the origin of social institutions. (ii) The account is no harder to apply than rival accounts. (iii) While reliance on counterfactuals cannot be fully avoided on this account, this reliance is both narrower in content and easier to justify. (iv) While the account needs to be amended with an individuation schema for social institutions and their features, the account is consistent with many such schemata, and should be seen to be required to itself provide (other than implicitly through standard inter-theoretic scientific bootstrapping).

For these reasons, presentist social functionalism can be seen to provide a compelling and fruitful characterization of institutional purpose, at least on the face of it. The next three chapters make clear that this impression is justified by considering three particular applications of the account. Apart from their inherent interest, these applications are also interesting, as they further spell out the responses to the four objections laid out here.

References

Acemoglu, D., S. Johnson, and J. A. Robinson. 2001. The Colonial Origins of Comparative Development: An Empirical Investigation. *The American Economic Review* 91 (5): 1369–1401.

Baum, D. A. 2013. Developmental Causation and the Problem of Homology. *Philosophy and Theory in Biology* 5.

Bertrand, M. 2013. Proper Environment and the SEP Account of Biological Function. *Synthese* 190 (9): 1503–1517.

Bigelow, J. C. 1998. Functionalism in Social Science. In *Routledge Encyclopedia of Philosophy*. Taylor and Francis.

Blake, C. F. 1994. Foot-Binding in Neo-Confucian China and the Appropriation of Female Labor. *Signs: Journal of Women in Culture and Society* 19 (3): 676–712. https://doi.org/10.1086/494917.

Brown, M. J. 2016. Footbinding, Industrialization, and Evolutionary Explanation. *Human Nature* 27 (4): 501–532. https://doi.org/10.1007/s12110-016-9268-5.

Darby, D. 2003. Feinberg and Martin on Human Rights. *Journal of Social Philosophy* 34 (2): 199–214.

Darby, D. 2004. Rights Externalism. *Philosophy and Phenomenological Research* 68 (3): 620–634.

Dawkins, R. 1982. *The Extended Phenotype*. Oxford: Oxford University Press.

Dawkins, R. 2004. Extended Phenotype – But Not Too Extended. A Reply to Laland. *Turner and Jablonka. Biology and Philosophy* 19:377–396.

Doolittle, W. F. 2014. Natural Selection Through Survival Alone, and the Possibility of Gaia. *Biology and Philosophy* 29:415–423. https://doi.org/10.1007/s10539-013-9384-0.

Dworkin, R. 1977. *Taking Rights Seriously*. Cambridge, MA: Harvard University Press.

Feinberg, J. 1992. *Freedom and Fulfillment: Philosophical Essays*. Princeton: Princeton University Press.

Fodor, J. 1990. *The Theory of Content*. Cambridge, MA: MIT Press.

Fresco, N., E. Jablonka, and S. Ginsburg. 2018. The Construction of Learned Information Through Selection Processes. In *The Routledge Handbook of Evolution and Philosophy*, ed. R. Joyce. Routledge.

Fuller, G., and A. Schulz. 2021. Idealizations and Partitions: A Defense of Robustness Analysis. *European Journal for the Philosophy of Science*. https://doi.org/10.1007/s13194-021-00428-8.

Futuyma, D. 2009. *Evolution*. 2nd ed. Sunderland, MA: Sinauer Associates.

Garrouste, P. 1994. Menger and Hayek on Institutions: Continuity and Discontinuity. *Journal of the History of Economic Thought* 16 (2): 270–291. https://doi.org/10.1017/S105383720000198X.
Gillespie, J. 1998. *Population Genetics: A Concise Guide*. 2nd ed. Baltimore: Johns Hopkins University Press.
Griffiths, P. 2005. Review of "Niche Construction". *Biology and Philosophy* 20:11–20.
Griffiths, P., and R. D. Gray. 1994. Developmental Systems and Evolutionary Explanation. *The Journal of Philosophy* 91 (6): 277–304.
Hayek, F. 1967. *Studies in Philosophy, Politics and Economics*. London: Routledge.
Hayek, F. 1973. *Laws Legislation and Liberty, Vol. 1: Rules and Order*. London: Routledge.
Kincaid, H. 1990. Assessing Functional Explanations in the Social Sciences. *PSA: Proceedings of the Biennial Meeting of the Philosophy of Science Association* 1990:341–354. http://www.jstor.org/stable/192715.
Ko, D. 1997. Bondage in Time: Footbinding and Fashion Theory. *Fashion Theory* 1 (1): 3–27. https://doi.org/10.2752/136270497779754552.
Ko, D. 2007. *Cinderella's Sisters: A Revisionist History of Footbinding*. Berkely: University of California Press.
Kusch, M. 2005. Fodor v. Kripke: Semantic Dispositionalism, Idealization and Ceteris Paribus Clauses. *Analysis* 65 (2): 156–163.
Laland, K. N., F. J. Odling-Smee, and M. W. Feldman. 2005. On the Breadth and Significance of Niche Construction. *Biology and Philosophy* 20:37–55.
Levins, R. 1966. The Strategy of Model Building in Population Biology. *American Scientist* 54 (4): 421–431.
Lewis, D. 1973. Causation. *The Journal of Philosophy* 70:556–567.
Mackie, G. 1996. Ending Footbinding and Infibulation: A Convention Account. *American Sociological Review* 61 (6): 999–1017.
Martin, R. 1993. *A System of Rights*. Oxford: Clarendon Press.
Merton, R. 1968. *Social Theory and Social Structure*. New York: Free Press.
Millikan, R. 1984. *Language, Thought, and Other Biological Categories*.
Millikan, R. 1990. Truth Rules, Hoverflies, and the Kripke-Wittgenstein Paradox. *The Philosophical Review* 99 (3): 323–353.
Millikan, R. 1996. On Swampkinds. *Mind and Language* 11 (1): 70–130.
Nanay, B. 2014. Teleosemantics without Etiology. *Philosophy of Science* 81 (5): 798–810.
Neander, K. 1996. Swampman Meets Swampcow. *Mind and Language* 11 (1): 70–130.
Neander, K. 2006. Content for Cognitive Science. In *Teleosemantics*, ed. G. F. Macdonald and D. Papineau. Oxford University Press.
Nickel, J. 1987. *Making Sense of Human Rights*. Berkeley: University of California Press.
Odenbaugh, J. 2006. The Strategy of "The Strategy of Model Building in Population Biology". *Biology and Philosophy* 21 (5): 607–621.
Odenbaugh, J., and A. Alexandrova. 2011. Buyer Beware: Robustness Analyses in Economics and Biology. *Biology and Philosophy* 26 (5): 757–771.
Odling-Smee, F. J., K. N. Laland, and M. W. Feldman. 2003. *Niche Construction: The Neglected Process in Evolution*. Princeton: Princeton University Press.
Orzack, S. H., and E. Sober. 1993. A Critical Assessment of Levins's The Strategy of Model Building in Population Biology (1966). *The Quarterly Review of Biology* 68 (4): 533–546.
Pettit, P. 1996. Functional Explanation and Virtual Selection. *The British Journal for the Philosophy of Science* 47 (2): 291–302.
Price, G. R. 1995. The Nature of Selection. *Journal of Theoretical Biology* 175:389–396.
Shepherd, J. R. 2018. *Footbinding as Fashion: Ethnicity, Labor, and Status in Traditional China*. Seattle: University of Washington Press.
Sober, E. 2008. *Evidence and Evolution*. Cambridge: Cambridge University Press.
Stalnaker, R. 1968. A Theory of Conditionals. In *Studies in Logical Theory*, ed. N. Rescher, 98–112. Basil Blackwell.

References

Tran, L. 2020. Footbinding as Fashion: Ethnicity, Labor, and Status in Traditional China. *The Journal of Interdisciplinary History* 50 (4): 632–633. https://doi.org/10.1162/jinh_r_01514.

Turner, D. 2007. *Making Prehistory: Historical Science and the Scientific Realism Debate*. Cambridge: Cambridge University Press.

Vaughn, K. 2013. Hayek, Equilibrium, and The Role of Institutions in Economic Order. *Critical Review* 25:473–496. https://doi.org/10.1080/08913811.2013.853862.

Weisberg, M. 2006. Forty Years of 'The Strategy': Levins on Model Building and Idealization. *Biology and Philosophy* 21 (5): 623–645. https://doi.org/10.1007/s10539-006-9051-9.

Open Access This chapter is licensed under the terms of the Creative Commons Attribution 4.0 International License (http://creativecommons.org/licenses/by/4.0/), which permits use, sharing, adaptation, distribution and reproduction in any medium or format, as long as you give appropriate credit to the original author(s) and the source, provide a link to the Creative Commons license and indicate if changes were made.

The images or other third party material in this chapter are included in the chapter's Creative Commons license, unless indicated otherwise in a credit line to the material. If material is not included in the chapter's Creative Commons license and your intended use is not permitted by statutory regulation or exceeds the permitted use, you will need to obtain permission directly from the copyright holder.

Chapter 4
Presentist Social Functionalism and the Function of Corporations

Abstract This chapter considers the question of the function of corporations. As is well known, two main answers to the question of what corporations are for have been proposed in the literature: the shareholder-value theory and the stakeholder-value theory. The chapter lays out these two views, and then argues that an appeal to presentist social functionalism shows that the terms of this debate are too constrained. In particular, by applying presentist social functionalism to this debate, it becomes clear that there are further possible functions of corporations that have not even been considered in the literature, and that existing studies of this question have looked at the wrong data, and thus fail to be able to resolve it. This chapter thus shows the practical applicability and fruitfulness of presentist social functionalism—especially compared to its rivals.

4.1 Introduction

Contemporary capitalist economic systems feature corporations—privately owned and independently managed entities dedicated to producing and selling a particular set of goods or services (Drucker 1993; Williamson 1981).[1] However, what is less obvious—and indeed a point of fierce contention—is what corporations are *for*. What is the *function* of corporations? Do they aim at maximizing payoffs for its shareholders only? Or do they aim at doing well for all of its stakeholders—including employees, customers, and all those affected by any externalities created by the corporation? There has been much debate about how to answer these questions (see e.g. Friedman 1970; Mulligan 1986; Freeman 1984; Orts and Strudler 2002).

---

[1] Drucker (1993, p. 4) has it that corporations sell goods "for profit" in a "competitive market." However, the former would beg the question in the present context, and the latter is overly restrictive, in that there can also be corporations in less than fully competitive markets (which are an ideal anyway).

Importantly—though this is less widely noted—in the background of these questions is the more general social scientific theoretical framework of functionalism. As noted in the previous chapters, determining what given social institutions are for allows us to get at important features of social reality; in turn, this can provide a fulcrum with which to understand, evaluate, and respond to this reality. It thus stands to reason that now that we have a compelling form of social functionalism at our disposal—presentist social functionalism—it will become possible to find new inroads into the question of the aims of corporations. This chapter seeks to support this impression.

In particular, as this chapter makes clear, this new, presentist social functionalist-based treatment of these issues is helpful for, on the one hand, making clearer what kind of data we need to collect in order to determine the function of corporations. On the other hand, the account shows that the function of corporations may be more complex than hitherto assumed and go beyond being focused just on shareholder benefits or just on stakeholder benefits. In this way, presentist social functionalism defended in this book can be shown to be a fruitful framework with which to assess the classic—but still important—question of the function of corporations.

Apart from its inherent interest, the discussion in this chapter is thus also important, as it further develops the account of presentist social functionalism laid out in the previous chapter. In particular, the discussion here further clarifies the inner workings of presentist social functionalism and makes explicit some of the benefits the latter has over alternative treatments of social functionalism.

The chapter is structured as follows. In sect. II, I set out the background of the dispute surrounding the aims of corporate activities in more detail. In sect. III, I apply presentist social functionalism to the question of the aims of corporate activity. In sect. IV, I contrast this approach to the alternative approaches in the literature and thus help make precise the benefits that presentist social functionalism has. I conclude in sect. V.

4.2 The Function of Corporations

A corporation is a privately owned and independently managed economic entity dedicated to producing and selling a particular set of goods or services (Drucker 1993; Williamson 1981). However, this very broad characterization leaves it open exactly what the aims of corporate activity are meant to be. Is it the case that the relevant set of goods and services is produced so as to advance the interests of those "owning" the corporation—i.e. its shareholders?[2] Or is it the case that the relevant set of goods and services is produced so as to advance the interests of all those having a significant relationship with the corporation—i.e. its stakeholders? Before it is

---

[2]The extent to which shareholding is equivalent to ownership is controversial (Bainbridge 2008; Stout 2002), but this is not so relevant here. See also below.

4.2 The Function of Corporations

possible to even begin to answer these questions, though, a few general remarks about their background need to be made.

First, it is of course true that anybody can make a corporation with any particular goal in mind. Now, it at first might seem that this pulls the rug out from under the entire debate surrounding the function of corporations. Since people's intentions in founding corporations can differ widely, there is no use in debating whether corporate activities should be seen to aim at shareholder benefits or stakeholder benefits. Both of these can be true—it just depends on the intentions behind the founding of the corporation in question.[3] In effect, this would be appealing to a kind of intentionalist, design-based account of social functionalism.

However, on a second look, it becomes clear that the fact that individuals or collectives can found corporations with many different goals in mind does not in fact spell an end of the debate surrounding the function of corporations. Even if we grant that this is one of the cases where individual intentions could ground the function of corporations—which, as noted in chap. 2, cannot be taken for granted, as this account is not general and there is no a priori to think it needs to apply here—given the complexity of corporations, there are good theoretical and empirical reasons for the thinking that this is at least not the *sole* way to ground their functions. The factors that determine whether and how corporations spread in a given socio-economic setting and whether they are stable parts of society—which, as noted throughout this book, are key aspects of social functional analysis (Bigelow 1998; Pettit 1996; Goodpaster 1991)—can differ from the intentions of the founders of a corporation. People might want their corporation to be stakeholder-focused—but corporate survival may depend on their ability to maximize shareholder value (or the reverse). This kind of issue will become crucial again in chap. 6, where it will be given a much more detailed treatment. For now, it is just sufficient to note that it is not obvious that the intentions of the designer are all that matters as far as the function of corporations is concerned—these intentions may be overridden by the "facts on the ground." Spelling out exactly how institutions obtain functions is the goal of the next section; however, what matters here is just that it is widely accepted that individual intentions are not the only way to ground the function of a social institution. This is sufficient for present purposes.

Something similar also holds for the idea that, in order to determine the appropriate aims of corporate activity, we simply consider the legal status of the relevant firm. Now, it is true that some legal systems have specific distinctions involving corporations. So, in the US legal context, limited liability corporations (LLC's) are distinguished, among others, from limited liability partnerships (LLP's), limited liability limited partnerships (LLLP's), and low profit limited liability corporations (L3C's) (Booth 2003; Artz et al. 2012; see also Segrestin et al. 2020). However, the existence of these legal distinctions and frameworks does not resolve the dispute concerning the functions of corporations.

---

[3] For a discussion of related issues in the context of recent changes to French corporate law, see Segrestin et al. (2020).

There has been a long debate surrounding the relationship between moral and legal demands, but all the major positions in that debate accept that existing laws need not match up, one-to-one, with social norms (Marmor and Sarch 2019). Legal distinctions are, in the first place, made for legal purposes, and need not match social—and certainly not moral—reality directly (Hart 1961; Raz 1979). While legal reasoning might need to be influenced by non-legal (and especially moral) considerations (Dworkin 1977, 1986), there is no reason to think that these non-legal considerations are the *only* thing that determines legal frameworks. The latter are also responsive to what is most useful or efficient for the legal organization of society (among other reasons).

So, it may be that corporations, from a social scientific perspective, have a given function F, but that we may want our legal codes to treat corporations in similar way to (say) consumer organizations with function F'. For example, we may want similar registration filing, tax reporting, compliance, and auditing procedures in both cases. If that is so, we may decide to create a *legal* category of "corporation" that includes both corporations proper and consumer organizations. However, the fact that the legal category of corporation includes consumer organizations with function F' and "proper" corporations with function F should not be seen to imply that corporations proper can have function F or F'. The latter is—by assumption—not the case. Alternatively, we may want to create a narrower category of legal corporation that includes all and only those proper corporations with function F and further features F' (e.g. being part of the Fortune 500). Again, this would then not entail that corporations not in the Fortune 500 are not corporations proper; it is just that, for legal reasons (e.g. having to do with the need for specific compliance requirements), *certain* "proper" corporations are given the different designation of being a *legal* corporation. Of course, we may often also want our legal categories to match social reality relatively closely. However, the key for present purposes is just that there is no requirement that this is so—and there are some good reasons to think it is frequently not the case.

All in all, therefore: the question of the function of corporations needs to be addressed head-on, and cannot be pushed aside as having a straightforward answer in either individual intentions or existing legal frameworks. A separate inquiry into this question is necessary.

To do this, it is best to begin by briefly reconsidering the two key views about the function of corporations: the shareholder-value theory and the stakeholder-value theory (see also Audi 2008). Note that these two are not the only views of the function of corporations in the literature.[4] However, they are the key poles around which

---

[4] In particular, Miller (2017, pp. 233–238; 2010, chaps. 2, 10) presents another account, according to which corporations have the function to provide "an adequate and sustainable supply of a good or service at a reasonable price and of reasonable quality" (Miller 2017, p. 231), and where the goods and services need to be defensible according to an objective moral standard (Miller 2017, pp. 235–236). However, while very interesting, this account will not be central in what follows. On the one hand, Miller's account has some internal problems. In particular, it is not clear how it is possible to specify "reasonable" prices and qualities of all the relevant goods and services, and it

4.2 The Function of Corporations

the debate turns, and most other positions are characterized in relation to them (see also Audi 2008).[5]

The first of these positions is often associated with Friedman (1970), but prominent defenses are also in Hansmann and Kraakman (2001), Stout (2002), and Jensen (2002) (among others). This position states that the function of corporations is to create benefits for their *shareholders*. The shareholders of a corporation are market entities (individuals or collectives) that provide funds to the corporation and receive a stake in the company that can be sold on (Bainbridge 2008; Stout 2002). Key reasons for thinking that this is what corporations are for is that the social value of the goods and services produced by a corporation is reflected in the value of its shares. Hence, increasing the value of the shares is bound to be correlated with the social value of the goods and services produced.

By contrast, the second position (which has seen prominent developments e.g. in Evan and Freeman 1988; Freeman and Reed 1983; Serafeim 2014) focuses on the benefits of the wider class of corporate *stakeholders*. The set of stakeholders of a corporation include the latter's shareholders as a proper subset, but has numerous other members as well: customers, employees, government regulators, competitors and those affected by any externalities produced by the corporation. Indeed, the stakeholders of a company are all those market entities with significant relationships to the corporation in question. A key motivating consideration behind the view that the function of corporations is the maximization of stakeholder benefits more generally is the fact that corporations affect many members of a society beyond its shareholders. For example, salaried employees, despite being crucial in producing the goods or services in question, need not be shareholders of the corporation they work for. The same goes for the corporation's customers, competitors, or those living near the places where the production takes place. While all of these social entities may be affected by the actions of the corporation, these effects need not be fully reflected in the value of the latter's shares. Hence, corporations should be seen to be acting in ways that take the interests of all of its stakeholders into account, not just those of its shareholders.

More details behind the reasons for these two views of the function of corporations could be given, but, for reasons made clearer in the next section, are not crucial in the present context. For now, it is best to use the machinery of presentist social functionalism and apply it to the case at hand, to see what progress can be made here.

---

is not clear how to determine which goods or services are objectively morally defensible (indeed, many scholars have argued that objective moral standards do not exist at all: Ayer 1936; Mackie 1977; Joyce 2001; Street 2006). On the other hand and most importantly, as will be made clearer below, it is not consistent with the most compelling ways of assigning functions to social institutions in the social sciences more generally. See also note 40 below and chap. 5.

[5] Also, the kinds of issues raised below can easily be expanded to the other theories.

4.3 Presentist Social Functionalism and the Function of Corporations

If it is neither the intentions behind the founding of the corporation nor its legal status that ground the function of the corporation, then what does? By the account of the previous chapter, this question can now be seen to turn on which features, in the current bio-cultural economic environment of a given capitalist system, increase a corporation's expected reproductive or survival success. So, corporations would have the function to maximize benefits for just its shareholders if doing so increases their expected survival or reproductive success (in the current economic climate) relative to those corporations that aim at maximizing benefits for all its stakeholders—and vice versa.[6] This reformulation of the debate surrounding the function of corporations is important, as it has two major implications that can significantly advance our understanding of corporations, and thus move this debate closer to a resolution.

First, the appeal to presentist social functionalism shows that prior treatments of the question of the function of corporations have relied on the wrong sort of data and inferences. Given this, these prior discussions—at least by themselves—can now be seen to in fact *fail* to shed light on the question of the function of corporations. (That said, as will be made clearer momentarily, they may well turn out to be useful *ingredients* in a compelling investigation of this question.)

For example, in the discussion surrounding the function of corporations, it is a common strategy to appeal to the values of the shares of shareholder-benefit-focused corporations in relation to those of stakeholder-benefit-focused ones (Hillman and Keim 2001; Jensen 2002; Stout 2002). (It turns out that stakeholder-focused firms do not clearly do better than firms that just focus on their shareholders: Hillman and Keim 2001; Jensen 2002; Stout 2002.)

However, given presentist social functionalism it becomes clear that these kinds of data do not in fact directly speak to the question they are meant to address. A corporation's stock market value is not the same as the probability that it survives or reproduces. A corporation's stock market value includes the value of the corporation's assets and people's expectations of the value of the company in the future. There is no question that these measures can correlate with the probability that the corporation survives or reproduces: corporations whose stock market valuation decreases are often in increased danger of going bankrupt. However, there is also no question that this correlation needs not always be high. Some corporations have high or increasing stock market values and go extinct and vice-versa. More generally, it is important to recognize that a corporation's stock market valuation is simply a different measure as that of its expected reproductive or survival success in the actually prevailing economic system, and they should not be conflated with each other. Because of this, considering a corporation's stock market valuation cannot be

---

[6] Note that which way of coding F—e.g. whether it is assigned to focus on shareholders only or all stakeholders—is arbitrary and does not affect the substance of the analysis that follows.

4.3 Presentist Social Functionalism and the Function of Corporations

directly used to determine its function—contrary to what is often assumed (Hillman and Keim 2001; Jensen 2002; Stout 2002).

Much the same is true for some other measures of whether the function of corporations is shareholder-benefit-focused or stakeholder-benefit-focused. For example, Bebchuk et al. (2021) and Bebchuk and Tallarita (2020) consider the actions of corporate leaders in US states that allow for the consideration of all corporate stakeholders in corporate transactions as compared to those of corporate leaders who have signed the Business Roundtable Statement on the Purpose of a Corporation (Business Roundtable 2020). The Business Roundtable Statement on the Purpose of a Corporation sees corporate leaders making a "fundamental commitment to all of our stakeholders" (underlining in original) (Business Roundtable 2020). Now, it turns out that these authors have found that corporate leaders hardly ever did, in fact, negotiate on behalf of all of their stakeholders during takeovers. Instead, they acted in ways that maximized the value for their shareholders (and themselves) only.

However, interesting as they are, these studies by themselves also fail to directly address the function of corporations. It is true that what corporate leaders do may, at least at times, be *related* to the relevant corporation's expected reproductive or survival success in the actually prevailing economic system. However, these are still very different measures. There are many influences on a corporation's expected reproductive or survival success in the actually prevailing economic system other the actions of corporate leaders, including the actions of suppliers, competitors, consumers, and employees. Because of this, it cannot be presumed that the former exhaust the latter. It may be true that corporate leaders only act in ways that favor corporate shareholders—however, it may also be true that doing so *lowers* the expected reproductive or survival success of corporations. In this case, then, corporate leaders act in ways that are *malfunctional*. The fact that a corporate leader does X does not mean that X is part of the function of a corporation. Hence, the consideration of the actions of corporate leaders alone cannot tell us what the function of corporations is. We have to look at the latter directly.

A related point holds for Friedman's (1970) classic concern that the CEO's of corporations focused on stakeholder-benefits are put in the position of supreme arbiters of different interests—something which they may lack a sufficient basis for. However, the fact that it may be difficult for CEO's to manage these different interests (or that many CEO's fail to do so well) does not speak to whether doing so is what is required of them in order to act in line with the function of corporations (Goodpaster 1991).

All in all: given the presentist functionalism defended in the previous section, the determination of the function of corporations needs to be based on measures of a corporation's expected reproductive or survival success in the current capitalist economic climate. While this implies that many of the currently available data are not useful to make this determination, it is now at least clearer what kinds of data we ought to be looking for—viz. data, concerning those features of corporations that increase their expected reproductive or persistence success. This thus makes for the first beneficial upshot of the defense of presentist social functionalism in relation to the question of the function of corporations: it brings into view the kinds of empirical investigations that need to be conducted so as to determine this function.

However, this does not exhaust the beneficial upshots of applying presentist social functionalism to the question of the function of corporations. The other major such benefit is theoretical. It concerns the fact that, with this account of functionalism in the background, we can explore novel theoretical possibilities concerning the function of corporations that have not even been *considered* thus far. In the forefront of these possibilities is the fact—already noted in the previous chapter—that functional ascription in the social sciences need not be seen to be restricted to one static feature of an institution only. That is, given presentist social functionalism, it becomes easier to investigate whether corporations have *neither* the function of maximizing benefits for its shareholders only, *nor* that of maximizing benefits for all of its stakeholders, but may have a more complex function.[7] Importantly, as also noted in the previous chapter, this functional pluralism would not be of the arbitrary kind, but be grounded in the socio-economic facts at hand.

In particular, given that the function of corporations is determined by those of their features that increase their expected reproductive or survival success, it now becomes clear that this function may depend on the features of both a focal corporation and those of *other* corporations. This also means that the function of corporations can change dynamically as the composition of corporations in the marketplace changes: that is to say, the features of a corporation that drive its reproductive or survival success—and thus ground its function—can vary over time and across socio-economic circumstances.[8] There are many different ways of spelling out these kinds of possibilities, but for present purposes, a very simple form of this kind of frequency-dependency is sufficient.[9]

Assume that, if rare in the relevant economic system, corporations that only maximize shareholder benefits have higher expected reproductive success than those focused on maximizing benefits for all stakeholders. The former produce highly sought-after goods and services at minimum cost. In turn, this might help them spread through the market: they are less likely to go bankrupt, and more likely to create offspring outlets (see also Schulz 2020). However, once these corporations are widely represented in the market, their fortunes may turn. Corporations that offer a wider set of benefits to a wider set of stakeholders—including employees,

---

[7] Note also that presentist social functionalism does not obviously support the fundamentally morally realist treatment in Miller (2017, pp. 233–238; 2010, chaps. 2, 10). There is no reason to think that corporations providing "an adequate and sustainable supply of a good or service at a reasonable price and of reasonable quality" (Miller 2017, p. 231), and where the goods and services are defensible according to an objective moral standard (Miller 2017, pp. 235–236), are the ones with the greatest expected reproductive success (see also Street 2006; Joyce 2005). See also note 37 and chap. 5.

[8] These features may continue to have similar effects (e.g. maximizing shareholder value). The issue is just that whether these features are functional—i.e. whether their effects contribute to the corporation's expected reproductive or survival success—will differ.

[9] Of course, in principle, the historical account can also allow for frequency-dependence. The problem, though, is that it is not compelling for other reasons, as made clear in chap. 2. See also below.

4.3 Presentist Social Functionalism and the Function of Corporations

customers, and the environment—will start to stand out as producing more valued goods and services. Despite their higher production costs, people are increasingly willing to purchase these goods and services, leading to the relevant corporations spreading through the market—and the cycle starting anew.

Formally, we might thus have it that the expected reproductive and survival success of the two types of corporations is as follows:

$$w_r = a_1 - bf_r \qquad (1)$$

$$w_s = a_2 + cf_r, \qquad (2)$$

where a_1, a_2, b, c, are all positive parameters, $a_1 > a_2$, w_r is the cultural fitness of shareholder-focused corporations, w_s is the cultural fitness of stakeholder-focused corporations, and f_r is the frequency of shareholder-focused corporations in the market.

In equilibrium, we have $w_r = w_s$, which implies that.

$$a_1 - bf_r = a_2 + cf_r = (a_1 - a_2)/(c+b) = f_r \qquad (3)$$

Depending on the details of the case, the upshot of this kind of scenario therefore is either a static stable state that contains both shareholder-focused and stakeholder-focused corporations, or a dynamic stable state that cycles through periods that contain mostly stakeholder-focused corporations and mostly shareholder-focused corporations (Orzack and Sober 1994). Graphically (Fig. 4.1):

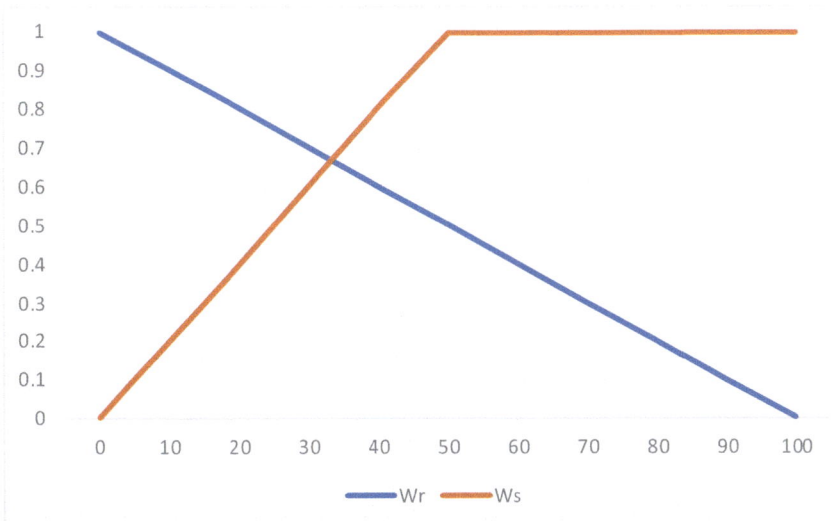

Fig. 4.1 Frequency-dependent cultural selection of corporations

As noted above, this is a highly simplified sketch of this case; however, what matters for present purposes is just that it illustrates a scenario in which it is not the case that the function of a corporation is to maximize either shareholder benefits or stakeholder benefits. Rather, something more complex is true here. The function of individual corporations differs from each other: in a mixed stable state, some corporations have the function to maximize just shareholder benefits and some the function to maximize stakeholder benefits. Dynamically, too, the above case sees the function of corporations—even of the same token corporation—as changing from maximizing just shareholder benefits to maximizing stakeholder benefits (and vice versa), depending on what happens in the wider economic system the corporation is part of. Generalizing, what this implies is that, on the above scenario, the best description of the function of corporations is dynamic and relativistic: this function depends on the details of the market environment—and which the corporations themselves are responsible for creating.

Now, it is important to be clear about the nature of the conclusion reached here. The claim is not that it is definitely true that there is frequency-dependency in the function of corporations. Rather, the key point here is just that this is a hypothesis that should be further explored. Up until now, the entire discussion surrounding the function of corporations was conducted in an either/or framework where corporations are either shareholder-benefit-maximizing or stakeholder-benefit-maximizing (or have some other unique feature F: see e.g. Miller 2010, 2017). However, given presentist social functionalism in the background, it becomes clear that this view of the universe of possibilities is too limited. There are many more options that need to be explored before it is possible to comfortably conclude what the function of corporation is. Indeed, it is thanks to presentist social functionalism that the dynamic and relativistic hypothesis concerning the function of corporations sketched here becomes even visible. This thus shows that presentist social functionalism is fruitful theoretical framework that can push existing the discussion surrounding the function of corporations into novel directions.

All of this matters further, as it brings into view a third benefit of the application of presentist social functionalism to the question of the function of corporations: namely, that it can advance the analysis and handling of corporate *corruption*.[10] The next chapter will consider institutional corruption more generally, but for now, it is sufficient to note some points specific to the case of corporations. In particular, consider the debate about whether it is defensible for big corporations to use tactics like persistent advertising meetings or biased information distribution to prevent employees from forming a union (Story 1995; Cooper and Patmore 2009). On the

---

[10] In line with the rest of the book, the focus here and in the next chapter is on institutionalist and functionalist treatments of corruption. There are also individualist, public choice-based accounts of corporate corruption (see e.g. Munger 2019). Importantly, though, these are consistent with what is presented in the text here: whatever public choice-based pressures there are on CEOs to engage in problematic, corruptive corporate activities, there is also the possibility that there are other activities—whether by a CEO or a corporation as a whole—that, while not individually corruptive, are institutionally so. Spelling this out is the goal here (and in the next chapter).

one hand, these tactics are often legal (Story 1995), and unionization is not necessarily something that benefits a corporation's shareholder value, its customers, or the wider economic system (see e.g. Cahuc et al. 2014, chap. 7, for a useful overview of the issues here; see also DiNardo and Lee 2004). On the other hand, unions can provide benefits to the employees of the corporation, and thus improve their well-being (Givan 2007; O'Mara 2019).

However, given presentist social functionalism, this question of the defensibility of anti-union activities of big corporations can now be investigated more precisely than what was possible beforehand. On the one hand, it is not implausible that a corporation's present expective reproductive or survival success may be increased by its fair, equitable, and honest dealings with its employees. If so, then benefiting its employees is part of the function of a corporation. If we further see activities that hinder an institution from fulfilling its function as *corruption*—which, as the next chapter makes clear, is not implausible—then anti-union action turns out to be a form of *corporate corruption*. However, whether this is in fact the case depends on whether it is actually—i.e. in the present socio-economic environment—true that a corporation's present expective reproductive or survival success is increased by its fair, equitable, and honest dealings with its employees, and whether unionization contributes to this.

Importantly, the latter may be true for some corporations at some time, but not for all corporations at all times. It is plausible that it will hold true in socio-economic systems where consumers are well-informed and have significant alternatives for the goods and services provided by a corporation. For in such cases, a corporation that does not treat its employees in a fair, equitable, and honest manner—which may (though need not) require allowing the formation of a union—may stand out negatively, and thus lose market share and be at a higher risk of bankruptcy or failure to spawn offspring firms (see also Schulz 2020, chap. 3). In these cases, therefore, treating its employees in a fair, equitable, and honest manner, is (at least ceteris paribus) part of the function of a corporation.[11]

This is important to note here for two reasons. First, the existing evaluations of anti-union action are typically done in an either/or manner: this either is, or is not, seen as defensible. What the application of presentist social functionalism here shows, though, is that the issues may be more complex: it may be defensible in certain cases or to a certain extent only.

Second, many of the existing evaluations of anti-union action are focused on the morality or justice of these actions—that is, whether they are defensible from the point of view of moral or political considerations. However, what the appeal to presentist social functionalism makes clear here is that this does not exhaust the kinds of considerations that are important to consider here. It is also relevant to assess how these actions relate to the function of corporations—i.e. whether they further or hinder the features that make it more likely for corporations to survive or reproduce.

---

[11] Of course, the same analysis can help clarify which union actions are corruption of the function of a union.

Importantly, these two sources of evaluation can complement each other. For example, to the extent that we seek to encourage corporations to treat their employees in a fair, equitable, and honest manner, this thus suggests regulating our socioeconomic system in such a way that corporation's expected reproductive or survival success is increased by its fair, equitable, and honest dealings with its employees. In turn, this may include ensuring that consumers *are* well-informed about the nature of corporate policies and actions (for example by encouraging the development of a free and widely read business press) and that they have many alternatives for the goods and services provided by a given corporation (thus encouraging the development of a robust competition regulator). Of course, there are many details to be worked out here about what exactly this entails. However, what matters for present purposes is just that the presentist social functionalist perspective defended here can make more precise exactly when, where, and why certain corporate behaviors (such as anti-union advertising) are corrupting.

Importantly also, the remarks of this section generalize to other social institutions. Whether the analysis is focused on privatized military contractors, campaign finance, policing (or whatever other social institution), a presentist functionalist perspective can (a) make more precise what data we need to consider in order to determine what the purpose is of the social institution in question (e.g. is the expected reproductive or survival success of a national defense force increased or decreased by its being managed by a private corporation or a public office?), (b) consider novel theoretical possibilities about this purpose (e.g. do elections for different offices have different functions—such as allowing the public to influence collective decision-making, as opposed to making people feel connected to each other), and (c) advance the analysis and handling of institutional corruption (e.g. would the installation of automatic speed cameras decrease the likelihood that the function of policing is undermined by the racial profiling of traffic violations?). For this reason, presentist social functionalism can be seen as a productive approach towards the study of institutional purpose and its corruption quite generally. Making this clearer is the aim of the next chapter.

4.4 Benefits of the Presentist Social Functionalist Analysis of Corporations

To further bring out the benefits of this way of approaching the function of corporations—and to deepen and illustrate the workings of presentist social functionalism more generally—it is now useful to contrast the treatment of the last section with the rival accounts laid out in chap. 2. Since the early discussion already engaged with the intentionalist-account, there is no need to go over these issues again. The focus in what follows will thus be on the historical, structural-functionalist, and virtual selectionist accounts.

4.4 Benefits of the Presentist Social Functionalist Analysis of Corporations

First, when it comes to the historical approaches, the key issue to consider is what features of corporations led to their cultural evolution: why did they spread? Alas, answering this question is quite difficult. On the one hand, it is not the case there were forms of corporations that were tried out, with some spreading rather than others. Of course, different countries have different frameworks for creating corporations, and within a country, the rules for creating firms tend to change across time. However, this is not well seen as a case where we can look to a history of past selection to determine the function of corporations: it is not the case that shareholder-focused firms were selected over stakeholder-focused firms; indeed, it is the origin of this dispute that both of these firms still exist.

On the other hand, it is not clear how far back we need to go to ground the function of corporations. Is the question which features led to the establishment of corporations in the early days of (Western) market economies? Or is the question which features led to the survival of corporations in the last 5 years (which include the Covid years)? Or is it both? It is not clear how to answer these questions, and thus, it is not clear how to use historical information to ground the function of corporations.

By contrast, the situation when it comes to presentist social functionalism is very different. As noted early, this allows for a clear—though complex and variable—set of functional ascriptions to corporations. Since functional ascription is now relativized to a specific time slice—the present—the ambiguities plaguing the historical accounts are absent. We can ask: here and now, do shareholder-focused firms do better—in terms of their expected reproductive survival or reproductive success—than stakeholder-focused ones? This means we can address this debate head on. We can also ask similar questions about other types of firms in the past: do family-run firms do better than non-family-run firms? Do firms arranged in guilds do better than those outside of guilds? As noted above, we can allow for these answers to differ in geographical and temporal contexts, and also allow for frequency-dependence and other complications. In this way, we can develop a sophisticated, dynamic account of corporate function that is still non-arbitrary and unambiguous. The function of corporations in country C_1 at time t_0 might be F_1, it might then change to a combination of F_2 and F_3 at time t_1—depending on the frequency of different firms of types A and B; all of this might further be different in country C_2. In this way, we can maintain the best of the historical accounts: the focus on the details of the historical situation of corporations in different cultures. Importantly, though, we obtain these benefits without having to countenance the drawbacks of the historical accounts: ambiguity and absence of the needed historical knowledge or facts.

Second, consider a structural-functionalist treatment of corporations (see e.g. Potts et al. 2016). This might take a number of different forms. For example, we may analyze the provision of goods and services to people, and in that context see corporations as playing the role—i.e. having the function of—ensuring an efficient production of these goods can be achieved. If so, we may favor a shareholder-based view of corporations. By contrast, if we consider a wider social system that includes people as providers of labor, consumers, family members, we may see corporations as playing the roles—i.e. having the functions of—providing goods and services to people, as well as providing them with stable employment, giving them access to important social networks, and safeguarding their and their families' futures. In turn, this might lead us to favor a stakeholder-based view of corporations.

For present purposes, the key point to note about this is that this makes the dispute surrounding the functions of corporations extremely difficult to resolve. In particular, this dispute now comes to look like the two sides are talking past each other: they are simply considering different questions. One side asks what role play for the provision of goods and services, the other asks what role they play in society more generally. To resolve this conflict, it is not a matter of collecting more or different data—e.g. about the performance of funds focused on ESG (environmental, social, and corporate governance) rather than more general funds. The issue is really what the right way of carving up society is here, and thus where to slot corporations into that analysis. Unfortunately, though, it is difficult to provide straightforward arguments for one or the other side on this point—this looks to be something that is inherently driven by the interests and dispositions of the relevant researchers.

The presentist social functionalist treatment of the previous section does not have this issue, though. Rather, it provides a non-arbitrary analysis of the function of corporations. The question is not which (social) system to analyze; rather the question is which features of corporations actually do increase their expected reproductive or persistence chances. As noted earlier, the presentist social functionalist treatment does point out that more and different data can help resolve this conflict: it is true that resolving this dispute does not turn on determining whether funds focused on ESG (environmental, social, and corporate governance) perform better than more general funds—we need data on which ways of treating employees (e.g. in terms of unionization) and customers (e.g. in terms of pricing) actually enhance a corporation's expected reproductive or survival success here. In this way, the dispute becomes non-arbitrary, resolvable, and not a case of the two parties talking past each other.

Finally, consider Pettit's (1996) virtual social functionalism. This would say that the function of corporations is tied to those of their features that ensure they would persist were their survival threatened. So, consider a world where our economies are drastically changed (due to some catastrophe or even by design) and we found ourselves in a state of much autarky, with many people being entirely self-reliant. If, in this counterfactual world, corporations persisted due to the fact that they bring different groups of people together to collaborate on solving collective action and provision problems, then we would be justified (at least ceteris paribus) in seeing corporations as having a stakeholder-like function. After all, it is in virtue of their ability to bring different groups of people together to collaborate on solving collective action and provision problems that corporations survive in cases where their existence would be threatened. By contrast, if in the above counterfactual scenario of a boost in autarkic living, corporations persisted due to their superior ability to create wealth for people, then we would be justified in seeing them as having a shareholder-like function (in the actual world). (Something similar for other functions, and even for the case where corporations have no functions, as they would not persist in the relevant scenarios where their survival were threatened.)

The problem with this approach here is that these kinds of counterfactuals are very difficult to assess, due to the fact that firms and corporations are so deeply embedded in market economies. On a general level, it is hard to imagine what the world would be like if the existence of firms or corporations were threatened. In a world where autarky was so common as to threaten the existence of firms, it is hard to say whether the latter would persist or why. This is akin to imagining a situation where Rawls's assumption of "moderate scarcity" does not obtain; in such a world, the precondition for cooperation do not seem satisfied. Would this collapse into a Hobbesian or Mohist state of nature? Or a more benign Lockean or even Rousseauian one? How do we know?

More generally, which threats to the existence of corporations should we consider here? Is the case of autarky sketched above too radical, as it already implies that corporations are unlikely to survive? If so, what alternative cases should we consider here? To be sure, as noted earlier, in a technical sense, corporations are quite culturally specific institutions. For example, they are not found among the Hadza people now, and did not exist even in Great Britain until relatively recently—much of the European Middle Ages saw production and consumption organized into small family-owned enterprises (smiths, bakeries, etc.). In that sense, it is easier to ask whether corporations would persist: if we introduced corporations into the world of the Hadza (as is increasingly happening: Apicella et al. 2014) or if we went back in time and introduced them into the England of Henry II, would they persist? However, even here, this kind of scenario is not fully tractable. These cases would require deep changes to our entire economy; it is hard to assess what would happen then. Perhaps the conditions would be right for the persistence of corporations, perhaps they would not be (Henrich 2020). This is likely to call for a detailed treatment of many specific contingencies of the case at hand. This is already very difficult for actual historical inquiries; assessing counterfactual situations in these ways is bound to be even harder.

This, though, is different when it comes to presentist social functionalism. Since the latter is based on the actual world only, we only need to assess what features of corporations (if any) aid their persistence or reproduction in the here and now. To be sure, as noted in the previous chapter, this kind of assessment can still be difficult. However, as also noted earlier in this chapter, this is not impossible. Indeed, presentist social functionalism points to the kind of data that can be collected to make these sort of inferences easier. This is a key upshot of the discussion in this chapter: it shows that the assessment of the features of an institution that increase its expected reproductive or persistence success are not nearly as shrouded in mystery as might be feared. In fact, as the discussion of the function of corporations shows, presentist social functionalism can make this assessment *easier*. In this way, the discussion here underwrites and illustrates the claim that presentist social functional avoids the pitfalls of Pettit's virtual social functionalism.[12]

---

[12] Note also that presentist social functionalism still allows for the possibility that corporations are institutions embedded into specific cultures that need not persist in all circumstances. However, the latter would be a consequence of the account, not its presuppositions: whether corporations would

In all of these ways, therefore, presentist social functionalism can provide a functional analysis of corporations that improves on that of its rival accounts. Indeed, it makes the question of the function of corporations tractable and avoids controversial (at best) historical, counterfactual, arbitrary, or moral claims.

4.5 Conclusion

This chapter developed a new account of the function of corporations. To do this, it applied presentist social functionalism to the question of the function of corporations. The upshot of this assessment is twofold. On the one hand, existing discussions of this question have tended to look at data that are not directly about the function of corporations. While stock market valuations and the behaviors of corporate leaders may sometimes correlate with the expected reproductive or survival success of corporations, they will not always do so, and are in general just a different measure. Instead, new studies should assess the expected reproductive or survival success of different types of corporations directly. On the other hand, the application of presentist social functionalism to the question of the function of corporations makes clear that there are numerous options about this function that have not even been considered thus far. In particular, this function may well be dynamic and relative in nature: different corporations, at different times, may be shareholder-benefit-maximizing or stakeholder-benefit-maximizing—or something else altogether. An appeal to presentist social functionalism can also make it clearer when and which corporate behaviors (such as anti-union advertising) are corruptive, and what regulatory changes may promote fair, equitable, and honest dealings between a corporation and its employees. The next chapter consider exactly this issue in more detail.

References

Apicella, C. L., E. M. Azevedo, N. A. Christakis, and J. H. Fowler. 2014. Evolutionary Origins of the Endowment Effect: Evidence from Hunter-Gatherers. *American Economic Review* 104 (6): 1793–1805. https://doi.org/10.1257/aer.104.6.1793.

Artz, N., J. Gramlich, and T. Porter. 2012. Low-profit Limited Liability Companies (L3Cs). *Journal of Public Affairs* 12 (3): 230–238. https://doi.org/10.1002/pa.1437.

Audi, R. 2008. *Business Ethics and Ethical Business*. Oxford: Oxford University Press.

Ayer, A. J. 1936. *Language, Truth and Logic*. London: ictor Gollancz.

Bainbridge, S. M. 2008. *The New Corporate Governance in Theory and Practice*. New York: Oxford University Press.

---

persist among the Hazda (say) is something we can use presentist social functionalism to help answer—it is not something we need to answer first in order to determine the function of corporations.

References

Bebchuk, L. A., Kastiel, K., & Tallarita, R. (2021). For Whom Corporate Leaders Bargain. *Southern California Law Review.* 94 (6). https://southerncalifornialawreview.com/2022/03/05/for-whom-corporate-leaders-bargain/

Bebchuk, L. A., & Tallarita, R. (2020). The Illusory Promise of Stakeholder Governance. *Cornell Law Review.* 106 (91): 91–178.

Bigelow, J. C. 1998. Functionalism in Social Science. In *Routledge Encyclopedia of Philosophy.* Taylor and Francis.

Booth, R. A. 2003. Form and Function in Business Organizations. *The Business Lawyer* 58 (4): 1433–1448.

Business Roundtable. (2020). *Statement on the Purpose of a Corporation* https://opportunity.businessroundtable.org/wp-content/uploads/2020/09/BRT-Statement-on-the-Purpose-of-a-Corporation-September-2020.pdf

Cahuc, P., S. Carcillo, and A. Zylberberg. 2014. *Labor Economics (W. McCuaig, Trans.; Second ed.).* Cambridge, MA: MIT Press.

Cooper, R., and G. Patmore. 2009. Private Detectives, Blacklists and Company Unions: Anti-Union Employer Strategy & Australian Labour History. *Labour History* 97:1–11.

DiNardo, J., and D. S. Lee. 2004. Economic Impacts of New Unionization on Private Sector Employers: 1984–2001*. *The Quarterly Journal of Economics* 119 (4): 1383–1441. https://doi.org/10.1162/0033553042476189.

Drucker, P. F. 1993. *Concept of the Corporation.* New Brunswick: Transaction Publishers.

Dworkin, R. 1977. *Taking Rights Seriously.* London: Duckworth.

Dworkin, R. 1986. *Law's Empire.* Cambridge, MA: Harvard University Press.

Evan, W. M., and R. E. Freeman. 1988. A Stakeholder Theory of the Modern Corporation: Kantian Capitalism. In *Ethical Theory and Business*, ed. T. L. Beauchamp and N. E. Bowie, 3rd ed., 97–106. Prentice-Hall.

Freeman, R. E. 1984. *Strategic Management: A Stakeholder Approach.* Boston: Pitman.

Freeman, R. E., and D. L. Reed. 1983. Stockholder and Stakeholders: A New Perspective on Corporate Governance. *California Management Review* 25 (3): 88–106.

Friedman, M. 1970. The Social Responsibility of Business is to Increase Its Profits. *The New York Times Magazine* 32-33:122–124.

Givan, R. K. 2007. Side by Side We Battle Onward? Representing Workers in Contemporary America. *British Journal of Industrial Relations* 45 (4): 829–855. https://doi.org/10.1111/j.1467-8543.2007.00663.x.

Goodpaster, K. E. 1991. Business Ethics and Stakeholder Analysis. *Business Ethics Quarterly* 1 (1): 53–73.

Hansmann, H., and R. Kraakman. 2001. The End of History for Corporate Law. *Georgetown Law Journal* 89 (2): 439–468.

Hart, H. L. A. 1961. *The Concept of Law.* Oxford: Clarendon Press.

Henrich, J. 2020. *The WEIRDest People in the World.* New York: Farrar, Straus and Giroux.

Hillman, A. J., and G. D. Keim. 2001. Shareholder Value, Stakeholder Management, and Social Issues: What's the Bottom Line? *Strategic Management Journal* 22 (2): 125–139.

Jensen, M. C. 2002. Value Maximization, Stakeholder Theory, and the Corporate Objective Function. *Business Ethics Quarterly* 12 (2): 235–256.

Joyce, R. 2001. *The Myth of Morality.* Cambridge: Cambridge University Press.

Joyce, R. 2005. *The Evolution of Morality.* Cambridge, MA: MIT Press.

Mackie, J. L. 1977. *Ethics: Inventing Right and Wrong.* Harmondsworth: Penguin.

Marmor, A., & Sarch, A. (2019). The Nature of Law. In E. N. Zalta (Ed.), *The Stanford Encyclopedia of Philosophy.*

Miller, S. 2010. *The Moral Foundations of Social Institutions.* Cambridge: Cambridge University Press.

Miller, S. 2017. *Institutional Corruption: A Study in Applied Philosophy.* Cambridge: Cambridge University Press.

Mulligan, T. 1986. A Critique of Milton Friedman's Essay "The Social Responsibility of Business Is to Increase Its Profits". *Journal of Business Ethics* 5:265–269.

Munger, M. 2019. *Is Capitalism Sustainable?* Great Barrington, MA: American Institute for Economic Research.

O'Mara, M. 2019. *The Code: Silicon Valley and the Remaking of America*. Newe York: Penguin Press.

Orts, E. W., and A. Strudler. 2002. The Ethical and Environmental Limits of Stakeholder Theory. *Business Ethics Quarterly* 12 (2): 215–233.

Orzack, S. H., and E. Sober. 1994. Optimality Models and the Test of Adaptationism. *The American Naturalist* 143 (3): 361–380. https://doi.org/10.2307/2462735.

Pettit, P. 1996. Functional Explanation and Virtual Selection. *The British Journal for the Philosophy of Science* 47 (2): 291–302.

Potts, R., K. Vella, A. Dale, and N. Sipe. 2016. Exploring the Usefulness of Structural–Functional Approaches to Analyse Governance of Planning Systems. *Planning Theory* 15 (2): 162–189.

Raz, J. 1979. *The Authority of Law*. Oxford: larendon Press.

Schulz, A. 2020. *Structure, Evidence, and Heuristic: Evolutionary Biology, Economics, and the Philosophy of their Relationship*. New York: Routledge.

Segrestin, B., A. Hatchuel, and K. Levillain. 2020. When the Law Distinguishes Between the Enterprise and the Corporation: The Case of the New French Law on Corporate Purpose. *Journal of Business Ethics*. 171:1–13. https://doi.org/10.1007/s10551-020-04439-y.

Serafeim, G. 2014. *The Role of the Corporation in Society: An Alternative View and Opportunities for Future Research*, 14–110. Harvard Business School Working Papers.

Story, A. 1995. Employer Speech, Union Representation Elections, and the First Amendment. *Berkeley Journal of Employment and Labor Law* 16 (2): 356–457.

Stout, L. A. 2002. Bad and Not-So-Bad Arguments for Shareholder Primacy. *Southern California Law Review* 75 (5): 1189–1210.

Street, S. 2006. A Darwinian Dilemma for Realist Theories of Value. *Philosophical Studies* 127:109–166.

Williamson, O. 1981. The Modern Corporation: Origins, Evolution, Attributes. *Journal of Economic Literature* 19 (4): 1537–1568.

Open Access This chapter is licensed under the terms of the Creative Commons Attribution 4.0 International License (http://creativecommons.org/licenses/by/4.0/), which permits use, sharing, adaptation, distribution and reproduction in any medium or format, as long as you give appropriate credit to the original author(s) and the source, provide a link to the Creative Commons license and indicate if changes were made.

The images or other third party material in this chapter are included in the chapter's Creative Commons license, unless indicated otherwise in a credit line to the material. If material is not included in the chapter's Creative Commons license and your intended use is not permitted by statutory regulation or exceeds the permitted use, you will need to obtain permission directly from the copyright holder.

Chapter 5
Institutional Corruption: The Presentist Social Functionalist Account

Abstract This chapter considers systemic corruption. Corruption is widely recognized to be a major social problem, but its characterization continues to be very controversial. Indeed, it is now commonly noted that what is being corrupted need not be an individual person at all but can be an entire social institution. This kind of institutional corruption has, especially in the last few years, come to be seen as ever more central and important. In this chapter, I advocate for a novel theory of this phenomenon, according to which it is the result of an individual or collective agent acting in ways that prevent a social institution from partially or fully fulfilling its function. In turn, the function of a social institution is spelled out in line with presentist social functionalism. The chapter shows that this new theory of institutional corruption is a useful addition to the literature, as it situates the study of this phenomenon in a wider functionalist approach toward the social sciences and does justice to the complexity of institutional corruption—both when it comes to its inherent nature and its moral evaluation.

5.1 Introduction

Corruption is widely recognized to be a major social problem, but its characterization continues to be very controversial. So, while it is frequently noted that corruption is "the abuse of power by a public official for private gain" (Nye 1967), not all corruption needs to involve public officials (doctors need not be public officials but can be corrupt if they prescribe medicine in accordance with who pays them to do so, rather than with what is best for the patient) or involve a private gain (when a county clerk grants wedding licenses in line with their personal moral or religious convictions and not the law, it can be a case of corruption but need not involve any private gain whatsoever).

Indeed, it is now commonly noted that what is being corrupted need not be an individual person at all but can be an entire social institution (Thompson 1995; Lessig 2013; Miller 2017; Ferretti 2018). This kind of institutional corruption has, especially in the last few years, come to be seen as ever more central and important (Thompson 2018; Miller 2017). Many of the major contemporary social problems appear to center on the undermining of institutions like voting, the free press, policing, or health care: instead of every citizen being equally able to influence political decision-making, to be informed about what is going on in the wider society, to be secure, or to be healthy, the institutions meant to provide these goods often seem to fail in their task (Thompson 1995; Lessig 2013; Miller 2017). This form of corruption thus deserves—and has seen—significant amounts of scrutiny in the last few years.

However, it continues to be a challenge to specify exactly what makes something a case of institutional corruption (Thompson 2018; Ferretti 2018). Exactly which actions subvert the relevant institution, and exactly why is it the case that these actions subvert the institution? What, specifically, is an institution's purpose? This chapter seeks to further the debate surrounding institutional corruption by answering these questions. After all, without a clear characterization of the nature of institutional corruption, fighting or avoiding it is difficult—for it is then not clear precisely what is to be fought or avoided (Rothstein and Varraich 2017).

The chapter, therefore, presents a general, philosophically and social scientifically well-grounded theory of institutional corruption that is centered on the idea that institutions have a social and not inherently normative function that is being subverted in cases of institutional corruption. While this theory shares some superficial components with some of the existing ones in the literature—especially those of Lessig (2013) and Miller (2017)—it is, in fact, quite different from the latter. In particular, by being built on the most compelling form of social functionalism—presentist social functionalism—the theory presented here has a solid theoretical foundation, does justice to the complex ethical nature of institutional corruption, and is in line with work in the social sciences more generally. Moreover, this theory is shown to have several important novel features: it is graded (institutions can be more or less corrupted), general (it can be applied to political contexts, but also many other social phenomena, from social media to private corporations and non-governmental organizations like the International Federation of Association Football [FIFA]), and unifying (it makes clear why highly corrupt societies tend to become unstable, whatever exactly the cause or moral status is of the corruption).

The chapter is structured as follows. Section I lays out the nature of institutional corruption and, following the lead of chap. 2, develops desiderata for its compelling characterization. Section II uses presentist social functionalism to develop a new non-normative teleological theory of institutional corruption that satisfies the desiderata of sect. I. Section III brings out some further novel benefits of the resulting treatment. Section IV concludes.

5.2 Institutional Corruption

Institutional corruption concerns cases where people engage in actions that undermine a particular social institution. These actions need not involve a private gain or *quid pro quo* exchanges of favor; indeed, these actions need not be inherently immoral or illegal. However, these actions still prevent the institution from operating as it is meant to. Such cases have come to be seen as being of major importance when it comes to ensuring that societies function in ways that benefit all their members (Miller 2017; Lessig 2013; Thompson 2018; Ferretti 2018; Ceva and Ferretti 2018; Della Porta and Vannucci 2012).

For example, in a given democracy, elections might be won only if candidates can obtain vast amounts of funding from major sponsors: only this ensures that they get heard or seen by voters. In that case, though, the only candidates who have a chance of obtaining office are those able to attract the necessary funds to finance their campaigns. This gives big political donors (businesses or wealthy individuals) an outsize influence on the running of the democracy. In turn, this can cause ordinary voters to feel like their voices do not matter, so they cease to participate in the political process. Thus, decisions are made in line with who can pay for access to these lawmakers, not who voted for them. At its extreme, this can spell the end of the relevant democracy. Similar points can be made about other examples, such as the privatization of prisons—which incentivizes incarceration rates and can thus decrease public security, in opposition to what prisons are for—and the mass dissemination of misleading or false information—which can undermine belief in public information of any kind (Satz 2013; Tsfati et al. 2020).

Cases like these have come to be seen to be of major importance: they are at the heart of some of the most widely discussed issues afflicting many contemporary societies (Satz 2013; Miller 2017; Lessig 2013; Thompson 2018). A number of theoretical proposals have been put forward to make the nature of institutional undermining that underlies them more precise (for helpful surveys, see e.g. Thompson 2018; Ferretti 2018; Brock 2014).

So, Thompson (1995) argues that institutional corruption concerns cases where public officials—especially legislators—receive political gains for providing services that are "procedurally improper" and which have a tendency to damage the political process (see also Philp 1997). Services are procedurally improper when they are not determined on the merits of the case, and/or they fail to follow the rules that ensure the political process is fair. If done systematically, such services can erode the public confidence in the political process—i.e., corrupt political institutions.

Not unrelatedly, Warren (2004, 2006, 2015) characterizes institutional corruption as instances where public officials *claim* to respect the egalitarian idea that all individuals affected by the collective decisions of the public officials should be able to influence these decisions, but where these officials *in fact* make their decisions so as to favor those who have provided benefits to these officials, and thus have privileged access to them. In other words, according to Warren's account, institutional corruption is at heart about duplicitous violations of democratic egalitarian ideals: public officials pretend to uphold these ideals, but do not actually do so, and that in a way that is in fact harmful to some members of the public.

There is no question that both of these characterizations of institutional corruption have allowed for many useful insights and advances. Most obviously, the problems caused by some forms of campaign finance for contemporary US democratic processes are well illuminated by both of these accounts: such campaign finance can be procedurally improper and in violation of egalitarian ideals of democratic political decision-making. Beyond this, the abstractness, especially of Warren's account, also makes clear what is wrong with other ills afflicting contemporary (representative) democracies, such as gerrymandering and voter suppression. These are cases that violate the egalitarian ideals at the heart of a genuine democracy—and they do that in a way that is surreptitious and thus hard to notice, avoid, and combat.

However, both of these proposals also struggle to go beyond this socio-political context and analyze institutional corruption more generally. It is not clear that these two proposals can be used to understand the institutional corruption of, say, prisons, the press, corporations, and not just that of political decision-making in representative democracies (and the US specifically). For example, the privatization of prisons is not obviously procedurally improper or done in a way that is democratically duplicitous. The issue with this privatization is not how it came about, which may have been entirely proper, or that it is inegalitarian, which it need not be, but that it undermines the institution it concerns. Much the same is true when it comes to the mass dissemination of misleading or false information (the source of which need not even be a public official at all). What matters is just that it concerns an undermining of the public press, not how it was decided on. In short: since institutional corruption is widely seen to comprise cases other than those of campaign donation in representative democracies, the proposals of Thompson and Warren appear insufficiently general—whatever other virtues they have.[1]

The account of Lessig (2013) is, therefore, a step in the right direction.[2] According to Lessig (2013, p. 553), "[i]nstitutional corruption is manifest when there is a systemic and strategic influence [...] that undermines the institution's effectiveness by diverting it from its purpose or weakening its ability to achieve its purpose." This influence need not be illegal, immoral, or procedurally improper; the key is just that it thwarts the *function* of the relevant institution. In this way, this account is significantly more general than the ones of Thompson and Warren. While it remains the case that the account in Lessig also tends to focus on the kind of ("dependence") corruption of the democratic political process that Thompson and Warren focus on, there is no reason that it cannot be easily extended to cover the corruption of the prison system, the press, and other public or even private institutions; indeed, it has

---

[1] See also Miller (2017, pp. 300–304). For a historical study of political corruption, see also Sparling (2019).
[2] The account is further developed in Lessig (2018).

been applied to the pharmaceutical industry with much success (see especially Lessig 2018; Fields 2013).

The main challenge the account faces is that it leaves open exactly what the function of a social institution is. What are prisons, or the press, or the National Collegiate Athletic Association (NCAA) for? Because of this, it also remains somewhat unclear exactly how this function can be undermined. Is the rise of social media undermining the press? Why? Without spelling this out, the account lacks a thorough theoretical grounding (see also Amit et al. 2017; Thompson 2018). Now, given that Lessig's focus also is the institutional corruption of the US political system—whose function may be relatively clear—this need not be greatly problematic for many of the uses Lessing has put his account to (see also Thompson 2018). However, as a full account of institutional corruption, Lessig's account falls short; while it has a sufficiently general overarching structure, this structure is not spelled out in enough detail to be able to make sense of institutional corruption in all of its different facets.

The account of Miller (2017) attempts to fill this lacuna. Like Lessig's, the account is teleological and general in nature; however, unlike that of Lessig, it is more fully spelled out.

According to Miller, social institutions are organizations—i.e., sets of structurally related functional roles (Miller 2017, p. 26)—that provide "collective goods by means of joint action" (Miller 2017, p. 23). That is, on this account, the purpose of a social institution is the provision, through the joint activity of the members of the institution, of objectively moral goods that are made available to all members of the relevant society (Miller 2017, p. 106). These goods comprise aggregated (needs-based) moral rights, freedoms, or well-being (Miller 2017, p. 23). Note that it is not sufficient that an organization provides collective goods that are *thought* to be moral goods; only organizations that provide collective goods that are *in fact* moral goods qualify as genuine social institutions (Miller 2017, pp. 23, 28, 34–45). In this way, the account of Miller (2017) makes it possible to provide a precise and systematic statement of what makes it the case that a given social institution has whatever function it has: namely, the fact that the collective intentions and actions of the members of the relevant society create institutions whose end is the obtaining of a collective, objectively moral human good. In turn, this also allows for a clear and general account of institutional corruption. Institutional corruption occurs when members of an institution intentionally engage in actions that tend to have the foreseeable and/or avoidable effect of undermining the function—spelled out as above—of the relevant institution (without, though, destroying that institution) (Miller 2017, pp. 82–88).

The account of Miller (2017)—like that of Lessig (2013)—is appropriately general. Since it makes the teleological nature of institutional corruption central to its characterization, it is not restricted or tied to the corruption of the democratic political process (important as that may be). Instead, it can be straightforwardly extended to other phenomena—such as the corruption of prisons or the press—for these two are instances where the provision of the collective moral goods (security and transparency) is thwarted (Miller 2017, pp. 217–218). Furthermore, the account of Miller

(2017) improves on the one of Lessig (2013), as it spells out this teleological nature of institutional corruption in detail (a point also noted by Thompson 2018). The function of social institutions is not left open as something to be filled in by the whims of the relevant researchers, but it is underwritten by a philosophically well-grounded treatment. However, Miller's (2017) account also faces three key drawbacks.

First, the theory does not speak to (what Miller calls) institutional corrosion (where actions are done that happen to slightly undermine the function of an institution but which fail the conditions for institutional corruption set out above), institutional destruction (where the institution is fully destroyed), or externally perpetrated institutional corruption (Miller 2017, pp. 66, 70). However, this restrictive focus of the analysis is not greatly compelling. Institutional corrosion, destruction, and external institutional corruption all lead to the same kind of failure of the provision of the relevant collective good as institutional corruption in its proper sense according to Miller (2017). While the source and exact nature of the prevention of the provision of the relevant collective moral good are different, the fact that there is this prevention is not. In this way, the account is overly limited. This is an important point to which I return below.

Second, the account of Miller (2017) needs to make strong commitments to highly contentious philosophical doctrines, such as a strong moral realism and methodological individualism. Now, it is of course true that it is not possible to settle large claims about moral realism and the like within the space of a book focused on an issue in social philosophy. However, what matters here is just that it is at least *far from clear* that these commitments are justified. For example, it is not obvious that the existence of objective moral facts—such as which collective goods are *in fact* morally good—can be made plausible (Mackie 1977; Joyce 2001; Street 2006). This makes it reasonable to see if there are other treatments of these issues that can avoid these commitments.

Third and most importantly, Miller's (2017) account is made problematic by the fact that it is fundamentally normative. On this account, institutional corruption *must* be morally bad (at least *pro tanto*): the moral appraisal of institutional corruption (and of social institutions in general) is *built into* the nature of institutional corruption (and institutions in general).[3] However, this fails to do justice to the moral complexity of institutional corruption (Lessig 2013; Thompson 2018). When it comes to the moral status of institutional corruption, everything depends on the details of the case and should not be built into the characterization of the nature of institutional corruption. People can engage in actions that lead to or constitute institutional corruption, but these actions can be morally neutral or even morally good (e.g., when the relevant social institutions are morally problematic).[4]

---

[3] Miller allows for the existence of "noble cause corruption," but this would be the case where corruption is engaged in for a (*pro tanto*) morally defensible reason: Miller (2017, chap. 4). However, this does not affect the main point in the text.

[4] This makes this different from some other related phenomena. For example, arguably, abusing one's power is always (*pro tanto*) morally bad: it concerns cases where a person acts against the

5.2 Institutional Corruption

Put differently, the normative focus of Miller's account makes this account arbitrarily limited. From the point of view of the underlying causal mechanisms—i.e., from the perspective of what is happening to the relevant institutions—the institutional corruption of the Mafia or the Nazi Party may be identical to that of US representative democracy or the press. While our normative evaluation of the former two cases may be different from the latter two, the social phenomenon underlying the four cases is the same: they share the crucial feature of thwarting the purpose of a social institution. They should thus be treated in the same way, too.[5] This is an important point to which I return in sect. III.

The point is further strengthened by the fact that the normative focus of Miller's account does not fit the long tradition of functional ascription in the social sciences more generally. As noted in chaps. 1 through 3, functional ascription in the social science is *not* normative in the way that Miller's (2017) account is. Importantly this is true independently of exactly which version of social functionalism we accept (including, in particular, presentist social functionalism). Of course, functional ascription in the social sciences is normative in the sense that all functional ascription is: it describes what a social institution ought to do, and not just what it, in fact, does do (Millikan 1984, 1989, 1990; Fodor 1990). Functional ascription in the social sciences may furthermore of course also be normative in the sense that that it says something which how social institutions contribute to the well-functioning of society and about how they do so—though this will depend on the account of social functionalism adopted (it is e.g. true for structural functionalism, but not necessarily on the other accounts laid out in chaps. 2 and 3).

However, functional ascription in the social science is *not* normative in the thickly moral sense that is presupposed by Miller's (2017) account. It does not rest on seeing social institutions as dedicated towards a specific set of objectively moral ends—partly for the reasons mentioned earlier that justifies what these ends would be and that they are, in fact, objectively moral has proven difficult. Hence, Miller's picture of functional ascription does not match that of the social sciences more generally—which is problematic, as the investigation of institutional corruption is a part of the social sciences.

Putting all of this together, it becomes clear that what is still needed is an account of institutional corruption that has the following three features:

---

reasons why they are in a position of power. It may be that a person aims at morally defensible outcomes by abusing their power, but the fact that they achieve these outcomes by abusing their power is one (moral) reason that speaks against doing so. However, this is different from cases of institutional corruption: the latter does not directly refer to ways of acting, but to the status of a social institution: *viz.* whether it is well-functioning. Put differently, an abuse of power *can result in* the corruption of an institution—but the latter can also result from behavior that is not an instance of the abuse of power. Importantly, also, since institutions can be morally good or bad, the well-functioning of these institutions can be morally good or bad as well. I thank Dale Dorsey for useful discussion of this issue.

[5] Miller considers the latter a case of "organizational corruption," and thus excludes it from the analysis: Miller (2017, p. 28).

1. *General*: The account needs to focus on the teleological nature of institutional corruption generally and not be restricted to the undermining of the (US) political process only.
2. *Spelled out*: The account needs to ground the function of social institutions in a plausible theoretical treatment and not leave it open to the intuitions of a researcher.
3. *Non-normative*: The account needs to spell out the function of social institutions in a way that does not presuppose that this function aims at some human good; rather, the moral valence of the social institution needs to be assessed depending on the details of the case.

An account that satisfies these three desiderata is able to combine the best features of the existing characterizations of institutional corruption while avoiding their drawbacks.

To make headway in developing such an account, the next section uses the framework of presentist social functionalism to lay out a novel account of institutional corruption that satisfies desiderata (1)–(3) and which has some further useful implications.

Before doing this, though, it is important to note that implicit in desiderata (1)–(3) is the idea that institutional corruption is, in its nature, quite different from individual corruption. As just noted, this is a common assumption in many views of institutional corruption (notably those of Thompson and Lessig), but it is not without controversy. For example, Ceva and Ferretti (2018) and Ferretti (2018) argue that institutional corruption reduces to individual corruption and that strongly separating out individual from institutional corruption obfuscates the mechanisms by which corruption spreads from one institutional context to another (see also Philp 1997). Relatedly, it is implicit in desiderata (1)–(3) that institutional corruption is to be analyzed teleologically (in terms of what the purpose is of the relevant social institutions) and not, say, deontologically (in terms of what it is our duty to do as members of a certain institution) or in terms of virtue (in terms of what virtuous members of the social institution are like) (see e.g. Rothstein and Varraich 2017).

Without question, there is a lot that could be said about these alternative, individualistic treatments of institutional corruption. However, instead of engaging in these debates directly, the approach here is the reverse. The chapter shows that adopting a teleological and non-individualist perspective on institutional corruption is coherent and has several advantages. In turn, this provides a reason for adopting this kind of view of institutional corruption. Of course, no pretense here is made that this has settled all the issues surrounding this issue (or, indeed, institutional corruption in general). Rather, the aim is more modest: it is just to show that a compelling, teleological, and non-individualist perspective on institutional corruption is available. If an alternative treatment is to be adopted, it would have to be shown to be superior while taking these benefits into account. (I return to these points below.)

With this in mind, consider applying presentist social functionalism to this case. As will be made clearer in the next two sections, this provides for a highly compelling basis for an account of institutional corruption.

5.3 Institutional Corruption: A Presentist Social Functionalist Account

With the presentist theory of social functionalism in the background, a novel characterization of institutional corruption can be developed that satisfies all of the desiderata laid out in sect. II and which has several further useful beneficial implications. In particular, given the plausibility of presentist social functionalism, institutional corruption can be characterized as follows:

> *Institutional corruption*: The extent to which the actions of a set of agents prevent a social institution N from fulfilling its function F, where F is the set of features of N that increase N's expected survival or reproductive success.

Several aspects of this characterization are important to note.

First, it is worthwhile making explicit how this characterization satisfies all of the desiderata laid out in sect. II.

> It is *general*: The present characterization of institutional corruption applies to all kinds of social institutions and is not restricted to the context of representative democracy (in the US or more broadly). This is due to the fact that the characterization is teleological and sees institutional corruption as the thwarting of the purpose of a social institution. Hence, it applies to any social institution with a function—which includes the prison system, the press, as well as the NCAA, corporations, or even such social institutions as the Mafia (among many others).

It is *spelled out*: The present characterization of institutional corruption is based on a well-grounded theory of the function of social institutions. Indeed, this is one of the two reasons why the defense of presentist social functionalism in chap. 3 is important here. This defense ensures that the characterization of the functional ascription of social institutions underlying institutional corruption has a strong theoretical basis and is not left to the intuitions of the relevant researchers.

It is *non-normative*: The present characterization of institutional corruption does not inherently see the purposes of social institutions as moral and, therefore, does not see institutional corruption as inherently normative. In this way, the present account of institutional corruption avoids the challenges faced by Miller's (2017) account: by making the ethical status of institutional corruption dependent on the details of the relevant institution, it can do justice to the ethical complexity of institutional corruption.

The fact that the above characterization of institutional corruption satisfies all of these desiderata further matters, as it shows that the notion of institutional purpose can be spelled out in a coherent manner and thus form the basis of a compelling account of institutional corruption. In this way, the present account can respond to some of the worries that have been levied against teleological accounts of this phenomenon more generally: namely, that its core notion—institutional purpose—cannot carry the weight it needs to (see e.g. Rothstein and Varraich 2017). As the defense of presentist social functionalism in chap. 3 makes clear, it *is* possible to provide a cogent grounding to the notion of institutional purpose and thus to use the latter as a foundation for a plausible account of institutional corruption.

This leads directly to the second important point to note about the present characterization of institutional corruption, which has also already been hinted at but deserves to be spelled out in more detail. This point concerns the fact that this characterization fits a general theoretical framework in the social sciences. This is the second major reason why the defense of presentist social functionalism from chap. 3 is important here. Unlike the account of Miller (2017)—which is also spelled out in detail—the present account is not disconnected from functionalism in the social sciences more generally. On the contrary, the present theory of institutional corruption is a natural extension of this general account of functionalism in the social sciences.

This not only gives this theory of institutional corruption a solid theoretical backing, but it also allows the easy extension of existing findings from the social sciences to the further investigation of institutional corruption. In particular, we do not need to establish the function of social institutions anew but can rely on the work already being done in the social sciences. For example, as also noted in chap. 4, we can rely on whatever theory of the function of corporations ends up being the most plausible one (whether it is the shareholder-benefit one or the stakeholder-benefit focused one), and we do not need to derive this function from scratch in the context of the investigation of potential institutional corruption. This way, we may also find instances of institutional corruption that we would have otherwise overlooked (for example, concerning the institutional corruption of corporations).

The third point to emphasize about this characterization of institutional corruption is that it does not require that the cause of the corruption is a systematic, intentional, immoral, or illegal action. Institutions can get accidentally corrupted, and they can get corrupted for moral or legal reasons. On the present account, institutional corruption is like the corruption of (electronic) data. If a flash drive (or printed out spreadsheet) falls into a river, it is likely that the data on it will become unusable and functionless. This is so whether the flash drive (or printed out spreadsheet) was intentionally, legally, or morally—or not—thrown into the river, and whether or not the data on the drive (or table) were moral or legal in content.

This is thus another way in which the present account does justice to the complexity of institutional corruption: it may sometimes require censure, it may be ethically problematic but excusable, it may be ethically neutral, and it may even be ethically permissible or even required. In this way, the present account can bring out what is common to all cases of the undermining of institutions (including corrosion, rebellion, and accidental prevention of function) without being forced to morally evaluate all of them in the same way. In turn, this places the normative and moral considerations squarely where they can do the most good: in the details of the relevant case.

For example, if someone acted in ways that undermined the function of the Nazi Party, then that may have been morally required. Indeed, even if this undermining of the Nazi Party is the result of mere laziness on the part of the relevant agent, it is still institutional corruption, and it is still morally good—though the person engaging in it need not deserve praise (Fricker 2016; Friedman 2013). For the same reason, the source of the corruption need not be systematic: just one action—such as

5.3 Institutional Corruption: A Presentist Social Functionalist Account

the distribution of fliers in front of the University of Munich—can (partly) undermine the Nazi Party and can thus count as institutional corruption.[6] (In a similar way, the data on a flash drive or printout can be corrupted with one-off behaviors—throwing it into a river—as well as with systematic actions, such as the careless treatment of the drive or piece of paper that, over time, leads to it getting dirty and unreadable.)

It is important to emphasize that the generality of the present account is one of its features, not one of its bugs. Of course, it is possible to make finer distinctions and focus particularly on certain forms of institutional corruption—say, ones that are internally, intentionally, and systematically caused and which target immoral institutions (as is done e.g. by Ferretti 2018; Miller 2010; Ceva and Ferretti 2018). However, this does not mean that there is not also value in providing a general account of the phenomenon. On the contrary, the generality of the present account is one of its key novel benefits.

In particular, by not using the sources and consequences of the undermining of an institution to characterize the nature of institutional corruption, it becomes possible to bring together what many superficially different social phenomena have in common. This is similar to generalizing accounts more generally. There are good reasons to often distinguish viral from bacterial diseases. However, there are also good reasons to often treat these subsets of the same overarching phenomenon: an infectious disease. This allows us to find common causes (e.g. the presence of other infected individuals) or common treatments (isolation, hydration, etc.).

In the present context, the fact that the account of institutional corruption defended here is generalizing allows us to note, for example, that some phenomena that might otherwise seem very different also share important communalities that it is theoretically and practically useful to note. For example, the Russia-based social media manipulation in the run-up to the 2016 US presidential election and the Trump administration's allegation of wide-scale voter fraud in the aftermath of the 2020 election differ in numerous particulars: the former is driven by sources external to US democratic institutions, the latter by sources internal to these institutions.[7] However, they also have some important features in common. In particular, they both (partially) prevented US democratic institutions from functioning properly, and they did so in similar ways—by increasing polarization and spreading propaganda.[8] This is theoretically valuable to bring out when studying democratic resiliency and the ways to improve it. For example, it suggests that similar responses may be useful in both cases, such as ensuring that the electorate is as well informed about the facts as possible. The fact that the present account of institutional corruption can bring out these communalities is thus one of its theoretical advantages.

---

[6] For this reason, Sophie and Hans Scholl can be praised for corrupting the Nazi party. (We can also praise someone for sabotaging—corrupting—a bomb so that it fails to go off and cause harm.)

[7] The two cases may also differ in intention and systematicity.

[8] In fact, this is shared with other cases, such as attempts to weaken the dictatorship in North Korea.

Similarly, it is a major benefit of the present account of institutional corruption that it brings out clearly that societies with many instances of institutional corruption are less likely to be stable. These are societies many of whose institutions are made less likely to survive or reproduce. Importantly, this is so independently of whether the corruption is systematic, intentional, or moral. On the present account, people living in highly corrupt societies—whatever distinguishing details there may be between these societies—have in common the fact that they need to deal with highly unstable institutions (i.e., institutions that face major barriers to their survival and reproduction). This brings out a key common feature of highly corrupt societies that other accounts would miss: whatever the details of their causes, a conglomeration of institutional corruption leads to institutional instability.

Importantly also, this is not a trivial inference. Rather, the present approach ties institutional corruption to the prevention of an institution fulfilling its function (and not to, say, duplicitous violations of democratic egalitarian ideals) and then spells out the function of an institution to those of its features that give it a current biocultural selective advantage. In this way, the present account can *explain* why societies with much institutional corruption are less likely to be stable—this follows from the present characterization of institutional corruption. Furthermore, this is not something that is, at least on the face of it, the case for the characterizations of Thompson (1995), Warren (2006), Lessig (2018), or Miller (2017), which would not lead us to expect much institutional corruption to go with much social instability: undemocratic and immoral societies can be stable.

Here, it is also noteworthy that not every crime or misdemeanor will count as an instance of institutional corruption on the present account. For example, ordinary theft need not block the function of an institution, and neither need all cases of nepotism (Miller 2017, pp. 110–115): the stealing of a bike need not have any implications for the institution of private property to survive or persist.[9] The present theory thus provides a general, encompassing account of the phenomenon without being either trivial or forced to accept contentious moral or metaethical propositions (as is true of other theories in the literature, such as that of Miller 2017). The present account allows us to separate the analysis of the presence and consequences of institutional corruption from its causes and moral status. This gives us more degrees of freedom in tackling this phenomenon in a way that is both feasible and compelling.

All in all, therefore, the present theory of institutional corruption sees it as the outcome of actions that partly or fully prevent a social institution from fulling its function—i.e., which partially or fully negate those features of the institutions that increase its expected reproductive or survival success. This theory is theoretically well-grounded in a compelling general account of social functionalism—presentist social functionalism—and it meets the desiderata of the previous section. It also has some further, novel benefits, as the next section makes clear.

---

[9] However, it is important to note that this will depend on the details of the case. If theft becomes sufficiently common, every additional theft could well make it harder for an institution of private property to persist. See also the discussion of graded institutional corruption in the next section.

5.4 Implications and Further Developments

This account of institutional corruption also has two further sets of benefits that make it stand out from rival accounts, and which make for an important addition to the literature. The first of these benefits is the fact that the source of the corruption need not be an individual human being but can also be a collective agent, like a corporation or foreign government. In particular, the characterization recognizes that an institution can be prevented from fulfilling its function by the concerted effort of a number of human beings.[10] For example, if a social network eases the spread of political misinformation, this can prevent the public press from fulfilling its function (see also Miller 2017, pp. 304–309). Importantly, this is so even if no individual can be seen as the source of this institutional corruption: owners and employees of the social network may not have been responsible themselves for furthering the spread of the misinformation—and may even have attempted to block this spread. Indeed, no individual user need have had any kind of significant impact on this spread. However, with sufficient numbers of users and sources of misinformation, misinformation can spread far and quickly, merely as the result of the structure of the institution of the social network (O'Connor and Weatherall 2021).

In this way, the present account diverges from those presented e.g. by Ferretti (2018), Ceva and Ferretti (2018), and Ferretti and Ceva (2021): institutional corruption need not reduce to the corruption of an individual agent.[11] To begin with—and as noted earlier—the institutional corruption need not be immoral, and even where it is, it need not result from the actions of a morally culpable individual. More importantly, though, the corruption need not even be analyzable into the intentions, ends, and behaviors of individual humans, as is assumed by Miller (2017). Rather, it can be the upshot of a genuinely collective agent (List and Pettit 2006). This matters, as it opens up a wider class of sources of institutional corruption and can thus help the study and prevention of the latter (see also Vergara 2020). In particular, the present characterization does not need to get involved in debates about the plausibility of individualism in the social sciences (see e.g. Elster 1982; Kincaid 2015; Jones 1996; Epstein 2014, 2015; Ruiz and Schulz 2023), but can work with whatever is the upshot of these debates. This is especially important due to the fact—noted earlier—that there is good reason to think that the holism/individualism debate may call for a pluralist solution that allows for both individualism and holism to sometimes be the best approach to a given social scientific issue (Ruiz and Schulz 2023). In this way, the present account's openness to collective agency and social holism

---

[10] This is a point also stressed by Miller (2017)—though, as noted below, the latter is committed to spelling out this kind of collective agency in individualist terms. See also Vergara (2020).

[11] As noted in the text and in the previous chapter, this is not to say that the present account needs to be an *opposed to* individualistic treatments of corruption—only that it goes beyond the latter. For more on individualistic, public-choice-focused accounts of corruption, see also Munger (2019a, 2019b); Coase (1960).

frees it from the constraints imposed by the individualistic commitments of Miller (2017), Ferretti (2018), Ceva and Ferretti (2018), and Ferretti and Ceva (2021).[12]

This deepens a point that was mentioned in sect. I already. Without a doubt, there is much complexity in the debate surrounding the question of whether all cases of institutional corruption reduce to cases of individual corruption. The same is true for the debate as to whether instead of a teleological account of the phenomenon, a deontological one (say) should be provided. The present treatment cannot be seen to address (or even to attempt to address) all the issues here. However, the point to note is that the present, teleological and non-individualistic account has several key benefits. In particular, it is coherent, it fits well to research in the social sciences elsewhere, and it brings out novel social patterns (such as the greater likelihood of instability in countries with many social institutions whose purposes are undermined). In turn, this means that good reasons need to be provided for giving up these benefits. If an individualistic, moral, and non-teleological account of institutional corruption is to be shown to be superior, it would have to be made clear that it has benefits, the sum of which is greater than that of the present account.

The second novel benefit to note about the account of institutional corruption defended here is that it is the first one in the literature that explicitly makes institutional corruption a matter of degree. This is important, as actions can prevent *some*, but not all, aspects of the function of a given social institution, and they can merely make the fulfillment of that function *harder*. For example, if one particular postal worker, out of tiredness, delivers mail a little late one day, then this is a very weak form of institutional corruption—if it is one at all: the function of the postal service is only negligibly undermined. By contrast, if postal workers are being so overworked—e.g., because of employment cuts—that they *all always* deliver mail late, then this is a more serious case of institutional corruption: the function of the postal service is seriously undermined. Finally, if the postmaster general orders the employees *not* to deliver mail, then that would be a very strong case of institutional corruption: the function of the postal service is fully undermined.[13]

The present account can easily handle this. It allows for institutional corruption to occur on a bigger or smaller scale: the greater the corruption, the more functions of an institution are undermined, and the more strongly they are undermined. The present account thus provides the right kind of framework with which to handle the complexity of the phenomenon. There is no need to make a call as to whether something definitely is or is not a case of institutional corruption; instead, we can allow something to be more or less of a case of institutional corruption. This is helpful, as existing accounts have tried to handle this fact by requiring genuine institutional corruption to be the result of actions that have the "tendency" to undermine the function of an institution (see e.g. Miller 2017; Thompson 1995). This, though, then

---

[12] Of course, this then raises a host of further questions concerning the ways in which collective agents can be morally responsible for their actions, etc. However, these questions can be left for a future occasion.

[13] Note also that these cases span different sources—individual actors and collective actors—as well as different degrees of systematicity and culpability.

5.5 Conclusion

requires an account of what such a tendency consists of and when it exists. In turn, this is not easy to do and may be somewhat arbitrary. It is clearer to describe the phenomenon as it is: namely, as leading to more or less of an undermining of the function of the relevant social institution. This is exactly what the present account does.

An example may make this clearer. Consider the International Federation of Association Football (FIFA).[14] This association may have a number of functions, including growing the sport of football internationally, advocating for fair play, and ensuring it is accessible to everyone. It has, however, been alleged that various actions have led to some of these functions being undermined; for example, its ability to advocate for fair play may have been hindered by some of its officials taking bribes for sponsorship contracts or the awarding of tournaments (see e.g. Jennings 2006). However, others of its functions—such as its ability to grow football internationally—may not have been so undermined. In this case, FIFA can now more clearly be stated to be *partially* institutionally corrupted, rather than us having to decide whether the actions of FIFA officials have, or have not, *fully* corrupted the organization.[15]

In short: the account of institutional corruption defended here has several further benefits—beyond the fact that it satisfies the desiderata of sect. II. In this way, it is especially good at doing justice to the inherent complexity of the phenomenon.

5.5 Conclusion

The characterization of and response to institutional corruption has come to be recognized as a major task of the social sciences (broadly understood). In this chapter, I advocate for a novel theory of this phenomenon, based on the framework of presentist social functionalism. According to this theory, institutional corruption is the result of an individual or collective agent acting in ways that prevent a social institution from partially or fully fulfilling its function. In turn, the function of a social institution is spelled out in line presentist social functionalism—i.e. as those of its features that increase its expected reproductive or survival success in the current sociocultural environment.

This theory of institutional corruption is a useful addition to the literature. It is teleological and thus general, fully spelled out, and non-normative. In particular, it ties institutional corruption to the thwarting of the purpose of a social institution and provides a solid theoretical grounding to these purposes, but it does not require them to be based on normative considerations. In this way, it situates the study of

---

[14] Another example are US campaign finance laws: instead of just considering individual corruption as a limitation on free speech and campaign finance, it becomes possible to consider some forms of campaign finance as being limited due to their partially systemically corrupting character (e.g. of the voting process). Further analysis of this, though, calls for a treatment of its own.

[15] Of course, these actions may also have been *individually* corrupt.

institutional corruption in a wider functionalist approach toward the social sciences and does justice to the complexity of institutional corruption—both when it comes to its inherent nature and its moral evaluation. Moreover, the discussion of this chapter also further brings out the fruitfulness of presentist social functionalism: it allows us to develop a new account of an important social phenomenon.

References

Amit, E., J. Koralnik, A.-C. Posten, M. Muethel, and L. Lessig. 2017. Institutional Corruption Revisited: Exploring Open Questions Within the Institutional Corruption Literature. *Southern California Interdisciplinary Law Journal.*
Brock, G. 2014. Institutional Integrity, Corruption, and Taxation. *Edmond J. Safra Working Papers 39.*
Ceva, E., and M. P. Ferretti. 2018. Political Corruption, Individual Behaviour and the Quality of Institutions. *Politics, Philosophy & Economics 17* (2): 216–231. https://doi.org/10.1177/1470594x17732067.
Coase, R. 1960. The Problem of Social Cost. *Journal of Law and Economics 3*:1–44.
Della Porta, D., and A. Vannucci. 2012. *The Hidden Order of Corruption: An Institutional Approach.* London: Routledge.
Elster, J. 1982. The Case for Methodological Individualism. *Theory and Society 11*:453–482.
Epstein, B. 2014. Why Macroeconomics Does Not Supervene on Microeconomics. *Journal of Economic Methodology 21* (1): 3–18.
Epstein, B. 2015. *The ant trap: Rebuilding the Foundations of the Social Sciences.* Oxford: Oxfor University Press.
Ferretti, M. P. 2018. A Taxonomy of Institutional Corruption. *Social Philosophy and Politics 35* (2): 242–263. https://doi.org/10.1017/S0265052519000086.
Ferretti, M. P., and E. Ceva. 2021. *Political Corruption: The Internal Enemy of Public Institutions.* Oxford: Oxford University Press.
Fields, G. 2013. Parallel Problems: Applying Institutional Corruption Analysis of Congress to Big Pharma. *The Journal of Law, Medicine & Ethics 41* (3): 556–560. https://doi.org/10.1111/jlme.12064.
Fodor, J. 1990. *The Theory of Content.* Cambridge, MA: MIT Press.
Fricker, M. 2016. What is the Point of Blame? A Paradigm-Based Explanation. *Noûs 50*:165–183.
Friedman, M. 2013. How to Blame People Responsibly. *The Journal of Value Inquiry 47*:271–284.
Jennings, A. 2006. *Foul!: The Secret World of Fifa: Bribes, Vote Rigging and Ticket Scandals.* London: Harper Sport.
Jones, T. 1996. Methodological Individualism in Proper Perspective. *Behavior and Philosophy 24* (2): 119–128. www.jstor.org/stable/27759348.
Joyce, R. 2001. *The Myth of Morality.* Cambridge: Cambridge University Press.
Kincaid, H. 2015. Open Empirical and Methodological Issues in the Individualism-Holism Debate. *Philosophy of Science 82* (5): 1127–1138.
Lessig, L. 2013. "Institutional Corruption" Defined. *Journal of Law, Medicine & Ethics 413*:553–555.
Lessig, L. 2018. *America, Compromised.* Chicago: University of Chicago Press.
List, C., and P. Pettit. 2006. Group Agency and Supervenience. *Southern Journal of Philosophy 44*:85–105.
Mackie, J. L. 1977. *Ethics: Inventing Right and Wrong.* Harmondsworth: Penguin.
Miller, S. 2010. *The Moral Foundations of Social Institutions.* Cambridge: Cambridge University Press.

References

Miller, S. 2017. *Institutional Corruption: A Study in Applied Philosophy*. Cambridge: Cambridge University Press.

Millikan, R. (1984). *Language, Thought, and Other Biological Categories*.

Millikan, R. 1989. Biosemantics. *Journal of Philosophy 86*:281–297.

Millikan, R. 1990. Truth Rules, Hoverflies, and the Kripke-Wittgenstein Paradox. *The Philosophical Review 99* (3): 323–353.

Munger, M. 2019a. *Is Capitalism Sustainable?* Great Barrington, MA: American Institute for Economic Research.

Munger, M. 2019b. Tullock and the Welfare Costs of Corruption: There is a "Political Coase Theorem". *Public Choice 181* (1): 83–100. https://doi.org/10.1007/s11127-018-0610-9.

Nye, J. S. 1967. Corruption and Political Development: A Cost-Benefit Analysis. *American Political Science Review 61* (2): 417–427. https://doi.org/10.2307/1953254.

O'Connor, C., & Weatherall, J. 2021. Modeling How False Beliefs Spread. In M. Hannon & J. de Ridder (Eds.), *The Routledge Handbook of Political Epistemology*. Routledge. 203–213.

Philp, M. 1997. Defining Political Corruption. *Political Studies 45* (3): 436–462. https://doi.org/10.1111/1467-9248.00090.

Rothstein, B., and A. Varraich. 2017. *Making Sense of Corruption*. Cambridge: Cambridge University Press.

Ruiz, N., and A. Schulz. 2023. Microfoundations and Methodology: A Complexity-Based Reconceptualization of the Debate. *British Journal for the Philosophy of Science 74* (2): 359–379.

Satz, D. 2013. Markets, Privatization and Corruption. *Social Research 80* (4): 993–1008.

Sparling, R. A. 2019. *Political Corruption: The Underside of Civic Morality*. Philadelphia: University of Pennsylvania Press.

Street, S. 2006. A Darwinian Dilemma for Realist Theories of Value. *Philosophical Studies* 127:109.

Thompson, D. F. 1995. *Ethics in Congress: From Individual to Institutional Corruption*. Washington, DC: Brookings Institution.

Thompson, D. F. 2018. Theories of Institutional Corruption. *Annual Review of Political Science* 21:495–513.

Tsfati, Y., H. G. Boomgaarden, J. Strömbäck, R. Vliegenthart, A. Damstra, and E. Lindgren. 2020. Causes and Consequences of Mainstream Media Dissemination of Fake News: Literature Review and Synthesis. *Annals of the International Communication Association 44* (2): 157–173. https://doi.org/10.1080/23808985.2020.1759443.

Vergara, C. 2020. *Systemic Corruption: Constitutional Ideas for an Anti-Oligarchic Republic*. Princeton: Princeton University Press.

Warren, M. E. 2004. What Does Corruption Mean in a Democracy? *American Journal of Political Science 48* (2): 328–343.

Warren, M. E. 2006. Political Corruption as Duplicitous Exclusion. *PS: Political Science & Politics 39* (4): 803–807.

Warren, M. E. 2015. The Meaning of Corruption in Democracies. In *Handbook of Political Corruption*, ed. P. M. Heywood, 42–55. Routledge.

Open Access This chapter is licensed under the terms of the Creative Commons Attribution 4.0 International License (http://creativecommons.org/licenses/by/4.0/), which permits use, sharing, adaptation, distribution and reproduction in any medium or format, as long as you give appropriate credit to the original author(s) and the source, provide a link to the Creative Commons license and indicate if changes were made.

The images or other third party material in this chapter are included in the chapter's Creative Commons license, unless indicated otherwise in a credit line to the material. If material is not included in the chapter's Creative Commons license and your intended use is not permitted by statutory regulation or exceeds the permitted use, you will need to obtain permission directly from the copyright holder.

Chapter 6
Artifacts: A Presentist Social Functionalist Account (Co-written with Clint Hurshman)

Abstract This chapter notes that assigning functions to artifacts and technologies is crucially important for a number of different reasons: in particular, it can explain how and why artifacts are used in the ways that they are, which artifacts are stable parts of society. In turn, this can help justify social policies and interventions. However, how to assign functions to artifacts is not yet fully clear. The most popular attempt to do so appeals to the intentions of the designer of the artifact—though others have tried to do so by appeal to the history of use of the artifact. However, in this chapter, we show that neither of these theories is fully compelling, and that a better account of artifact function can obtained by applying a variant of presentist social functionalism to this case. We lay out this presentist theory of artifact functions and apply it to two examples of controversial artifact functions: off-label uses of pharmaceuticals, and policy responses to ChatGPT and other large-language models.

6.1 Introduction

Other than its particular social nature, a key aspect of human life is the fact that it is characterized by its heavy reliance on tools and artifacts. We humans are not just dependent on others to make our way through the world, but we are also building technology to aid us in this. As it turns out—and as will be made clearer later in this chapter—there are some important connections between the social nature of human living and its technological underpinnings. However, even intrinsically, assigning functions to artifacts is very important.

First, functional ascriptions can be used to *describe and identify* the artifact. A hammer is defined by what it is for: to drive nails into objects (say). Lots of things can be used, more or less well, *as* hammers (sticks, screwdrivers, flutes, keyboards), but that does not mean that they *are* hammers. Only objects that have the function to drive nails into objects are actually hammers.

Second and most importantly, functional ascriptions to artifacts play important social scientific roles. On the one hand, they can explain why certain artifacts are stable parts of society, and others not. The function of music tapes was to record and play music, not to be book weights. This explains why they went out of existence when better ways of distributing music became available, and were not affected by the existence of better book weights. On the other hand, artifact functions can also be used to justify interventions. If X is for F, then its failing to do F justifies fixing it so that it does F. A hammer that fails to drive nails into objects needs fixing. A lightbulb that has the same "problem" does not need fixing.

However, it is not yet fully clear how to justify a particular functional ascription to an artifact. As this chapter (together with the previous ones) makes clear, the two main ways to do so—the intentionalist and the historical selectionist accounts—have problems that mean these accounts are not fully compelling as they stand. This is particularly important when it comes to the intentionalist account, which is central in the case of artifacts, and is thus especially well illuminated here.

To alleviate this situation, the chapter applies presentist social functionalism to this case. This not only provides a more compelling way to assign functions to artifacts, but, by making the connections between artifacts and social institutions clearer, it brings out novel aspects of the discussion surrounding technology. It also again clearly displays the fruitfulness of the presentist social functionalist perspective.

The chapter is structured as follows. In sect. II, we present the two major existing accounts of artifact functions and bring out their problems. In sect. III, we develop the presentist social functionalist alternative and note its benefits. In sect. IV, we apply this presentist social functionalist treatment of artifacts to some contemporary cases: ChatGPT (and other large language models) and off-label uses of pharmaceuticals. Section V concludes.

6.2 Existing Accounts and their Problems

Artifacts bear the mark of intentional design, and much of the study of artifacts has emphasized this fact—elaborating, for example, the ways that intentions are communicated to users (Dipert 1993, 1995; see also Norman 2013) and the role that intentions may play in individuating artifact kinds (Evnine 2016). It is often taken as self-evident that talk of the functions of technology must refer to the intentions behind its design: "No agent, no purpose, no function" (McLaughlin 2000, p. 60). Intentions therefore form the basis of the prevailing theory of artifact functions: the *intentionalist theory* (Vermaas and Houkes 2003).

Intentionalist theories (see e.g. Dipert 1993, 1995; Hilpinen 1993; McLaughlin 2000; Houkes and Vermaas 2004, 2010; Vermaas and Houkes 2003) take intentions of designers or users to be *necessary* to justify proper function ascriptions to artifacts. In this sense, intentionalist accounts are often described as "agent-oriented" rather than "property-oriented" (Van Eck and Weber 2014, p. 1371).[1]

One might also take a designer's intention to be *sufficient* for an artifact to have a concomitant function. Most views, however, put further conditions on justified function ascriptions, which have to do with the ways that artifacts are epistemically situated. Houkes and Vermaas (2010), for example, suggest a *support requirement* and a *communicative requirement*. Designers are assumed to *understand* the artifacts they create more deeply than most users, and to *communicate* uses to users, facilitating a social division of knowledge. For Houkes and Vermaas (2010), in order for a function ascription to be justified, a designer must have a *support belief*, according to which an artifact has the physical properties necessary to be at least minimally capable of performing the function that they assign to it. They can thereby justify a *use plan*, a set of instructions to produce the functional effects with the artifact.

The communicative requirement also contributes to this social division of labor. If the intentions of designers occupy such a privileged position, then users must have some way of knowing them. Functions can be communicated by word of mouth, advertising, instruction manuals, and other media. They can even be signaled by the artifact itself—for example, an ergonomic grip on a hammer can signal how it is meant to be held.[2] However this communication takes place, it is often taken to be a core aspect of artifact functions, since it maintains the epistemic (understanding-focused) structure described above. Indeed, Dipert (1995) explicitly requires that artifacts intentionally signal their functions. Houkes and Vermaas (2010) likewise require that functions originate from the aforementioned "justifiers"; if one is not a justifier herself then she must rely on communications from other users.

In some cases, this communication will be superfluous, namely when one makes an artifact for one's own use. Suppose I create a cake that looks like a plant pot. This artifact would not signal its function to users—in fact, it may misleadingly signal functions that it is not capable of performing. Yet it still seems capable of performing *its* function—that of a cake—when I eat it. For Houkes and Vermaas (2010), communication will not be necessary because I occupy the epistemic position of a justifier, at least at the time of baking. Nevertheless, even here there may be an epistemic asymmetry between designer and user. At the time of baking, I must know a recipe and various practices associated with baking. I need not know any of these things at the time of consumption. I am "permitted" to occupy a different epistemic position as a user than that which I must occupy as a maker. If communication is unnecessary here, it is because of my epistemic position as the maker. In general, however, communication is critical for artifact use, since it allows agents to use

---

[1] Likewise, Houkes and Vermaas (2010) describe their view as a "function-ascription formulation" rather than a "function-as-property formulation" (p. 78).
[2] Relatedly, Norman (2013) suggests principles of design that are meant to help users infer correct use as easily as possible. See also Davis (2020).

artifacts according to their purposes without fully understanding *how* they achieve those purposes.

Similarly, users often use artifacts in nonstandard ways, which may vary in their effectiveness. A novel use does not count as a proper function unless a user has sufficient knowledge about an artifact to rationally justify her new use plan in terms of the artifact's physical properties. Houkes and Vermaas (2010) call such users *justifiers* and suggest that they participate in the artifact's design since they occupy an epistemic position analogous to that of designers and from this position "design" uses for the artifact. Artifacts are characterized by the social division of epistemic labor that is thus achieved (Houkes 2006), whereby "passive users" (Houkes and Vermaas 2010) can make use of artifacts without understanding the mechanisms by which artifacts work. If this is meant to expand to other tokens of the artifact, though, these novel use intentions need to be communicated to others—hence the need for the communication requirement.

In short: on the predominant theory of artifact functions—the intentionalist one—an artifact type A has function F if the designers of the artifact (which may include its users) (i) intend to A to do F, (ii) believe A can do F, and (iii) communicate the intention for A to do F to users. While this theory has many compelling features, it also faces two major challenges.

First, designers' intentions are often indeterminate. People can and do simply intend to build or fashion something that seems useful for a lot of different things—some of which they may not even be aware of yet. I can pick up a stick and sharpen one end, simply because I think it might come in handy for a variety of uses. I don't intend it for any specific use, but I think it might turn out to be useful as a spear, or a hammer, or a walking stick. The same is true for contemporary artifacts like smartwatches, which also were designed with a variety of causal powers with little inherent communalities—tracking health and fitness data (heart rates, etc.), relaying messages, playing music, telling the time. No specific use-cases need to have been in the minds of the designers when they designed these watches; their intention may just have been to build an electronic device that can be worn on the wrist and that has a variety of causal powers.

If so, though, then the intentionalist-account would have to say that these artifacts do not have a determinate function. This is not obviously the right answer, though: for example, it may turn out that most people end up using smartwatches as fitness trackers. As noted in the discussion of the function of corporations, this might then make it social scientifically reasonable to see fitness tracking as the function of these watches.[3] We return to this issue in more detail in the next section, but for now, the key point to note is just that the fact that a designer's intentions are often unclear makes them a weak basis on which to build an account of artifact functions.

---

[3] It may be tempting to say that this constitutes a new act of design (as sketched earlier). However (as also noted earlier), this would then fall foul of the communication requirement of the intentionalist account: users need not broadcast their intention to use smartwatches as fitness trackers.

6.2 Existing Accounts and their Problems

The second problem with the intentionalist account of artifact functions is the fact that patterns of use at least sometimes seem to be able to override intentions (see also Preston 1998, 2009, 2013; Scheele 2006). This is again related to some of the arguments made in the chaps. 4 and 5, and will be made clearer below. However, for now, it is sufficient to note that a maker may design an artifact A to do F, but people mostly or always use it do G. This could be because people's values or intentions diverge from those of the designers, or because they have discovered uses of the artifact not recognized by the designers.[4] In that case, though, it seems that the role that functional ascription here plays, social scientifically, would seem to be focus on G. For example, frisbees were designed as pie containers, but came to be mostly used as toys, WD-40 was designed as a rust-preventer, but it is mostly used as a mechanical lubricant. Saying that frisbees are being mis-used as toys or WD-40 is being mis-used as a lubricant would run counter to these established practices and is unlikely to be socially scientifically compelling.

In the background of this objection is the need for an account of what it is to design or create an artifact. Is picking up a stick creating a hammer? If so, what if I start using the stick as a fishing spear? Is that equivalent to creating a new artifact? Houkes and Vermaas (2010) have an account of function change that occurs through use. For them, functional ascriptions must be traceable back to designers or "justifiers": creative users who understand the properties of artifacts well enough to support novel uses. For Houkes and Vermaas, functions change when justifiers participate in the "design" of a new kind of artifact; for example, those who used Frisbie pie containers as flying discs thereby designed a toy that used the pie container as material. Once these designers could *support* their use plan with an understanding of the artifacts and *communicated* this use plan to others, the function changed. Below, we will argue that our account provides a better explanation of function change than this, at least for our purposes. For now, it suffices to say that the emphasis on the intentions of designers, or indeed on those of any privileged epistemic agents, sits oddly with the flexibility of artifact use.

Indeed, this is related to the point that was made in the previous chapters, and which will also become important again below: a key role that functional ascription plays in the social sciences is to help us determine which parts of science are stable, long-terms features of that society, and which merely transient phenomena. In this context, if an artifact is a stable feature of society—as underwritten by its pattern of use centering on G—then focusing on the designer's intentions for A to do F would miss this. Again, this will be made clearer in the next section below, but for now, the key idea is just that the intentions of a designer, even where they are determinate, do not clearly seem to be a compelling basis for a functional ascription in cases where patterns of use diverge from these intentions.

Consider next the other major account of the artifact functions: the historical, reproduction-based account. The structural-functionalist and virtual selectionist

---

[4] As before, this does not amount to the creation of a new artifact, given that the communication requirement here fails.

accounts are not central in this discussion, and since they were discussed at length already, will not be re-considered here. In fact, even when it comes to the historical account, the discussion here can be quite brief, as several of the main points have already been made in the previous chapters. Still, a few remarks specifically in the context of functional ascription to artifacts are important to make.

Numerous authors have compared the development of technologies to evolution by natural selection (Ziman 2000). Biological functions are valuable because they are *explanatorily deep*: they give insightful explanations for the *existence* of characters to which they are ascribed (Garson 2019; see also Wright 1976). Similar explanations are possible for many cultural entities, of which artifacts are just one type (Boyd and Richerson 1985). A theory of artifact functions that takes this approach can be called a *reproduction* theory. The most developed reproduction-based view is given by Preston (1998, 2009, 2013), who draws primarily from Millikan's (1984) theory of proper functions in biology. Preston (2013) gives the following definition of artifact functions:

> A current token of an item of material culture has the proper function of producing an effect of a given type just in case producing this effect (whether it actually does so or not) contributes to the best explanation of the patterns of use to which past tokens of this type of item have been put, and which in turn have contributed to the reproduction of such items. (Preston 2013, p. 187)

As in the biological case, functions are meant to explain differential reproduction resulting in currently existing tokens. In particular, Preston appeals to social processes to explain patterns of use, and these same patterns of use can help to explain the intentions. A watch found on a beach, for example, indicates not just a designer who created it to satisfy a specific intention, but almost certainly an entire cultural history that itself explains why a designer would intend to keep time in the first place (Preston 2013, pp. 157–158). Designers' intentions are just one part of a larger explanatory picture. Even when they play a prominent role in the explanations of the patterns of use of artifacts, however, artifacts do not have their functions in virtue of being designed with specific intentions, but in virtue of being explicable by a kind of selection.

Incorporating non-intentional factors into the functions of technologies may also be fruitful because non-intentional aspects of technologies have arguably been those of most interest to philosophers of technology for decades. Winner (1980), in his seminal essay on the politics of artifacts, compares technologies, in their power to restrict human freedom and regulate behavior, to "legislative acts or political foundings that establish a framework for public order that will endure over many generations" (p. 128). He uses nuclear power to illustrate this restrictive aspect of technology (see also Mumford 1964). Nuclear power, Winner (1980) argues, is predicated on institutions that make it safe to use; once the technology is in use, dissolving the institutions that oversee production and handle radioactive byproducts becomes a practical impossibility. Regardless of whether this assessment of nuclear power is correct, the point is that a critical part of what technologies *do* to individuals and societies is independent of what designers and users intend to do

6.2 Existing Accounts and their Problems

with them. Technologies are more than instruments for our use. A non-intentionalist account like Preston's opens up the possibility of incorporating these political footprints into the functions of technologies.

However, there are also several problems for this way of ascribing functions to artifacts. On the one hand, as before, many technologies seem to lack the necessary history to make this kind of functional ascription possible. For example, it may be social scientifically useful to ascribe functions to Facebook, even before there was a history of social media platforms to fall back on. These issues are familiar from the rest of this book, and so do not need to be rehearsed here. However, there is also a more specific second problem for the account that needs to be discussed here.

This second problem concerns the fact that the notion of selection at the heart of the account is not fully clear. As formulated by Preston (1998), functional ascription to an artifact depends on whether appealing to effect E *contributes to* the *best explanation* of the patterns of use to which past tokens of this type of item have been put, and which in turn have *contributed to* the reproduction of such items. However, there are a few underspecified aspects of this characterization. It is not entirely clear *how much* of a contribution to the best explanation the effect has to have—or exactly what this *best* explanation consists in. Nor is it clear what contribution to the reproduction of the artifact the relevant patterns of use are meant to make. For example, it is not clear whether it matters that the patterns of use incidentally increase (or decrease?) the reproduction of the artifact because they are *correlated* with something else that affects this reproduction. In general, it is not clear exactly what is meant by the "reproduction" of these artifacts. In short, the relationship between an artifact, its features, its patterns of usage, and its "reproduction" is not fully clear on this account.[5]

All in all therefore, the intentional account has the benefit of avoiding an appeal to the past history of the artifact or unclear statements about its patterns of use and reproduction. However, it is problematic for focusing on the designers' mental states—which are also often highly unclear—and for failing to pay attention to patterns of use, which do seem to be very important for functional ascriptions to

---

[5] Another issue that is sometimes discussed here concerns "phantom functions" (see Holm 2017). A phantom function "occurs when a type of item is regularly reproduced to serve a specific function, but no exemplar of it has ever been structurally capable of performing that function" (Preston 2013, p. 138). For example, we might say that an amulet has the function to ward off ghosts—though it cannot actually do that in a world without ghosts. However, both the intentionalist and the reproduction-focused accounts can, in some form, account for these. For example, Houkes and Vermaas (2010) could note that such artifacts violate the aforementioned support requirement by being unable to perform their functions. The function ascription is therefore unjustified—but phantom functions nevertheless have much in common with genuine functions. Similarly, Preston (1998) could note that, since artifact functions, like biological functions, must be successfully performed sometimes (see also Walsh and Ariew 1996), phantom functions cannot be genuine functions—although the artifacts under consideration might have other functions. For example, while an amulet never actually wards off ghosts, it might make one feel safe by making her *think* that it is warding off ghosts. Hence, the appeal to phantom functions, while interesting for other reasons, is not a good way to assess these theories here. This point will be picked up again later in this chapter.

artifacts. By contrast, the historical, reproduction-focused account is good for focusing precisely on these patterns of use and avoiding appeal to the designers' mental states. However, it is burdened with a historical focus and an unclear account of the role of patterns of use on the reproduction of the artifact. By applying presentist social functionalism to this case, we can get the best of both of these explanations—while still avoiding their pitfalls. The next section makes this clearer.

6.3 The Presentist Social Functionalist Account

To develop a new, presentist social functionalist treatment of artifact functions, the first thing to do is to build on the insight of the historical, reproduction-based account that the best way to analyze artifact functions is by placing them into the context of their pattern of use—that is, to see artifacts and social institutions as closely intertwined.

Technologies and institutions are continuous. As noted in the discussion of Winner (1980) in the previous section, artifacts can affect the world and shape institutions in ways that do not depend on the intentions of users or designers. This continuity between artifacts and institutions at the level of the power they exercise makes it reasonable to try to understand their functions under a common framework. Latour (1990) also argued that technology is "society made durable," by which he meant that the social world cannot be stable without the presence of non-human actors; society necessarily involves the interplay of both human agents and material things. Technology is in part constitutive of the social, and the social is also in part constitutive of technology (Naoe 2008). Social processes play a part in determining where (and whether) boundaries exist between artifact kinds (Feenberg 1999, chap. 4; Bijker and Pinch 1987).

Most importantly for the present chapter, artifacts and institutions also seem to be related in the sense that we want their functions to perform some of the same discursive work. As noted throughout the book, it is important for social scientists to tell us where a society is going (Bigelow 1998)—in part for its own sake, but also so that we can intervene to affect the future of that society. A principled way of identifying institutional corruption is useful, at least in part, because undesired effects resulting from corruption seem to demand different interventions than undesired effects resulting from an institution that is functioning well. For example, debates about whether the unjust effects of policing demand reform or abolition might be read as hinging on how we identify the *function* of policing. If presentist social functionalism yields a useful notion of institutional failure—as argued in the previous chapter—then this provides some reason to think that an analogous view will yield a useful notion of technological (mal-)function.

With this in mind, consider how to apply presentist social functionalism to the case of artifacts. Consider an artifact A that is being used in a variety of ways F_1 to F_n. For example, this could be a "hammer" that is used to insert nails into walls, to

6.3 The Presentist Social Functionalist Account

hold books in place, measure the straightness of a board, smash rocks, accessorize clothing, etc. Then, we propose that:

> Artifact type A has function F_i (out of F_1 to F_n) if usage pattern F_i of tokens of A increases the expected reproductive or survival success of the type A in the current social environment.

To understand this better, several points need to be made.

First, as before, this account is presentist, actualist, and non-arbitrary, and thus avoids the problems of the historical or structural functionalist accounts. This is in much the same way as was the case for the other applications noted in the book, and so there is no need to discuss this further here.

Second, though, the account is also non-intentionalist. In particular, it avoids the problem of the intentionalist account for being focused on the—often unclear— intentions of an artifact's designers. Indeed, it is also focused on the patterns of use of the artifact, rather than the artifact creation. To understand this better, it is important to note the role that the type/token distinction plays in the above account. The goal of the present account is to assign functions to artifact *types* (Franssen 2006; Franssen et al. 2014; Evnine 2016) by considering the effects of patterns of usage of their tokens.[6] In this way, artifact tokens can be seen to get their functions as a result of their membership in kinds.

This distinction matters, as the persistence or reproduction of a type does not always entail the persistence of a token, and vice-versa. Using a hammer as a fashion accessory may increase the survival chance of individual token hammers, but not (it may be presumed) that of hammer types: the type hammer is not likely not stay in existence because of that fact that individual hammers are being used as fashion accessories. No new hammers are produced, and old hammers will increasingly break or get lost. By contrast, using hammers to drive nails into walls is likely to lead to the production of more hammer types, even if it leads to the destruction of individual hammer tokens. For a more extreme version of this sort of case, consider batteries. The use of a specific token battery does not promote the continued use of that token battery beyond the short term, since it is used up in the process. However, it may promote the continued use of batteries in general: if they prove useful for a given purpose, thereby satisfying a user's intention, they are more likely to be used for that purpose in the future. They may also promote the persistence of their kind by non-intentional means by promoting the use of devices that will require more batteries in the future. Thus, by performing its function, the token contributes to the stability of its type.

---

[6] Franssen (2006) uses a slightly different terminology, and distinguishes artifact kinds from artifact tokens. This is somewhat non-standard, though: in metaphysics, *kinds* are typically said to have *members*, while *types* have *tokens*. However, Franssen uses artifact "kind" as a technical term, which he describes as a variety of artifact "type" (though Franssen also makes use of a narrower notion of artifact *type*) (p. 48). Thus, kinds have tokens, for Franssen, because they are a *type* in the sense typically used in metaphysics. However, here, we are using the standard terminology of type / token. These are just verbal issues, and if, desired, Franssen's alternative terminology could be substituted for the one used here.

This is not to say that artifact types and tokens never line up in their persistence conditions. If my car gets me to work, I'm more likely to keep using it than I would be if it failed to do so. If my car is a very good one, I may want to use it as long as possible; in this sense, the car's functioning may contribute to its own persistence. This is one way of contributing to the persistence of its type; continuing to use my car is sufficient for continuing to use *cars*. In many paradigmatic cases, however, artifacts contribute to the stability of their types in broader ways that extend beyond the continued use of a given token. My car affords uses that cause me to depend on cars *in general*; it allows me to live far from where I work and even in the absence of public transportation, such that I depend on having a car to survive. As a result, I may be more disposed to buy a new car after my current one stops working than I would be if my way of life had not been so shaped by driving. The point is that even when artifact tokens contribute to their own persistence, by shaping users and societies, they typically contribute to their persistence of their types more broadly, as well; however, even contributing to the persistence of a single token is sufficient for contributing to the persistence of a type. In sum: the explanandum of artifact functions on the persistence theory is the stability of artifact *types*.

This also matters—and this is the third key point to note here—in that it helps to make the relevant notion of "reproduction" clearer. As noted earlier, this was left unclear in the key historical, reproduction-based account. By contrast, the idea here is clearer, and avoids appeal to "best explanations" or vague notions of "contribution." Rather, the claim is just this: if using a hammer in a certain way makes it more likely for the type hammer to stay in existence or to lead to the reproduction of new, offspring types of hammers, then that way of using hammers is part of the function of hammers. In this way, the account allows for both reproduction and persistence, and as made clearer in chaps. 2 and 3, fills a lacuna left by the historical account.

The fourth point to note here is that, in the background of this account of artifact functions is some way of classifying artifacts into types. This is parallel to the issue of the individuation of institutions and their features noted in chap. 3, though, and thus does not need to be rehearsed here. Also, as was the case there, this again allows for a bootstrapping treatment: an individuation schema for artifact types can, at least in part, be justified though its usefulness in a functionalist treatment of artifacts.

In short, on the present treatment, the function of an artifact is not readable off of the artifact itself, nor off of its design, but depends ineliminably on the social context in which the artifact is used. Artifact types acquire functions through those patterns of use of their tokens that increase their (i.e. the types') expected reproductive or persistence success. Importantly, this account can also account for the six hallmark features of artifacts that Preston (2009) suggests any adequate account of artifact functions should be able to address: multiple realizability, multiple utilizability, recycling, reproduction with variation, malfunction, and phantom functions. Considering these features is useful for making the nature of the present account clearer.

6.3.1 Multiple Realizability

Artifact functions, in general, are multiply realizable. This means that objects with different physical properties can perform the same function, at least in principle. Corkscrews can be made out of a variety of materials, lightbulbs can use LEDS or wires, etc.—and such differences need not imply different functions.[7] At the root of this is the fact that, typically, not everything an artifact does is its function, so its non-functional properties can be changed without changing the function. However, it is not always clear which features are functional and which are not, so it is important that a theory of artifact functions can help draw the distinction. The present account of artifact functions does this, and thus allows for, and indeed explains, multiple realizability.

To see this, consider any of the above examples. So, different kinds of corkscrews and different kinds of lightbulbs seem to persist for largely the same reasons (respectively), even when the relevant kinds are made in different ways.[8] While some things that an artifact does promote the persistence of its kind, other things it does will be irrelevant. The latter features can vary across artifacts that nevertheless have the same function.

6.3.2 Multiple Utilizability

Artifacts are also multiply utilizable, in that they can be put to different uses. Because not all of an artifact's properties are relevant to its function, any given artifact could likely serve a variety of purposes, including ones that are not its proper function. Thus a spoon can be used like a shovel to dig holes. Digging does not thereby become its function, since it presumably plays no part, or at least a marginal part, in explaining why spoons persist. If, however, agents began to routinely use spoons to dig holes, such that digging became a significant part of the explanation for why spoons persist—in other words, a significant part of what spoons are *depended on* to do—then the function of spoons would change.

In other cases, materially identical artifacts may have different functions because they are members of artifact kinds that are stable for different reasons. To build on the above example, suppose the type of spoon used in one culture is materially identical to a type of small shovel used in another culture. In the two cultures, the artifacts promote their persistence by producing different effects, and in the face of different selection pressures. Therefore, they form distinct functional types. This is

---

[7] There is much discussion of how to best understand multiple realizability (see e.g. Polger and Shapiro 2016; Shapiro 2004; Weiskopf 2011). However, these differences do not matter here.
[8] There might, nevertheless, be fine-grained distinctions within kinds. Wooden and metal spoons, for example, share some persistence conditions, and do not share others. This is true on other accounts as well, however (see Franssen 2006).

possible because functions do not supervene on physicochemical properties. Identical objects can therefore count as artifacts of different kinds. We need not invoke the intentions of users or designers to explain this, however. Rather, we can focus on the ways that artifacts are used and the roles that they play in society, which need not be intended or acknowledged (see also Lachney and Dotson 2018).

6.3.3 Recycling

Recycling is another hallmark feature of artifacts that follows from multiple realizability and multiple utilizability. It results from the facts that any given artifact has properties relevant to a variety of possible functions, and can thus be re-purposed. Now, when Preston (2009, 2013) considers recycling, she assumes that recycled artifacts are modified in some way to serve their new purposes. If an artifact is simply used for a non-standard purpose without modifying it in any way, that purpose is only a system function.

However, this move is untenable, as non-standard uses for artifacts often become standard without any such modification. There is nothing infelicitous about calling a mason jar used exclusively for drinking a drinking glass, or calling a tire a couch. Modification could be a part of the change in an artifact's proper function, but in many cases it is unnecessary. The above, presentist-social-functionalist-based account can easily handle this fact: all that matters is that the artifact type is used in different ways than it has been before. Moreover, this account also fits to and improves on Scheele's (2006) account of function ascription to artifacts. Scheele (2006) considers situation of the "Pieterskerk"—a gothic church in Leiden in the Netherlands that has served a variety of purposes over the course of centuries. Originally intended for religious purposes, it has effectively ceased to be used for religious purposes at all and is now used as an event space. Scheele writes:

> Although in most cases the proper functions of artefacts are determined by the designer or manufacturer this is not necessarily the case. An artefact that is used by everyone in a way alternative to the intentions of the designer will very soon change its proper function, due to the fact that this new use will become generally accepted. The reason for this is, roughly, that the socially accepted use will have changed, i.e. the relevant social norms will have changed and thus have overruled the original function ascription. (Scheele 2006, p. 30)

Scheele therefore argues that the function of the Pieterskerk has changed because the consensus about it has changed; after secularization had long led people to treat the Pieterskerk as something other than a church, it was finally bought by a private organization for use on a "quasi-commercial basis" (Scheele 2006, p. 29).

The present account shares with that of Scheele the fact that it deemphasizes the epistemic authority of designers. However, unlike Scheele (2006), the present account of artifact functions does not appeal to the social consensus surrounding the artifact's function—Scheele's view remains agent-oriented in a way that the present account is not. This is beneficial for two reasons. First, as before, the determination of when a "social consensus" is reached is difficult. Avoiding appeal to the latter thus makes the present account clearer. Second and most importantly, the present account has the benefit—shared with the treatment of institutional corruption of the previous chapter—that it is gradualist. Suppose that at t_1, "spoons" are strictly used for eating foods, and using them as such increases their expected reproductive or survival success. Now suppose that at t_2, "spoons" are used solely for digging holes, and using them as such increases their expected reproductive or survival success. The function of spoons will have thus changed: they have turned from utensils to gardening tools (say). Importantly, this change may (though need not) have occurred gradually: for much of their history, spoons may have been used both for eating and digging. The present account can easily allow for this: the ability to be used for digging may have increased the reproduction of the type "spoon" initially not all, then somewhat, then fully determined it. This is likely to be true in many cases of recycling, where artifacts lend themselves to different uses that each contribute to the persistence of some artifact kind in different ways. In this way, the present account can account for the complex, dynamic, and gradualist nature of artifact functions, much as was true for the function of corporations as laid out in chap. 4.

As noted above, the intentionalist theory also has an explanation for function change. In order for an improvised use to become a proper function, on the intentionalist theory, it must be communicated by a justifier who can support the new use with beliefs about the physicochemical properties of the artifact relevant to performing the new function. This requirement is appropriate if a goal of a function theory is to explain the rational justification of artifact use. As noted above, the intentionalist theory involves a social division of knowledge that explains why it is rationally justifiable for agents to use artifacts that appear to them as black boxes (Houkes 2006). However, from the perspective of social science, it is sufficient to *explain* why agents use artifacts in the ways that they do, even if this explanation does not amount to a justification. Thus, for our purposes, it is preferable to relax this epistemic requirement.

6.3.4 *Reproduction with Variation*

Preston (2009) laments that the literature on artifact functions, specifically on the intentionalist side, has focused heavily on *production* to the neglect of *reproduction*. It is clear that this is not an issue that affects the present account. By serving users' intentions, artifacts encourage their continued use; alternatively, by creating habits or dependencies in users, artifacts can seem to *demand* their continued use. However, an artifact kind's conditions of persistence may change over time as its selection pressures change. Both reproduction and variation can be explained by pointing to the ways that artifacts are used and are socially situated, as is done on the presentist social functionalist account defended here.

6.3.5 Malfunction

Another important phenomenon for a theory of artifact functions to explain is *malfunction*: how is it that artifacts can have functions that they are not performing? An account must do more than simply describe what an entity is doing in order to capture this normative aspect of functions. As noted in the previous chapters, malfunction is an important phenomenon for any functionalist theory to describe. The presentist social functionalist account of artifact functions defended here defines function in terms of the present contribution of tokens to the persistence of their type. An artifact malfunctions, then, when it fails to be useable in ways that would promote the persistence/reproduction of its type, given the socio-cultural selection pressures it currently faces. If my car's engine breaks down and I am unable to drive it, it is malfunctioning: the ability to drive them is what underlies the spread of cars (it may be assumed). However, if my car's engine gets muffled, and I am thus unable to make loud noises with it, it is not malfunctioning: making loud noises is not what drives the spread of cars (it may be presumed). This is exactly parallel to the treatment of institutional corruption offered in the previous chapter.

6.3.6 Phantom Functions

Lastly, an account of artifact functions should have something to say about *phantom functions* (Preston 1998, 2009, 2013; Holm 2017). A phantom function is one that is ascribed to an artifact that it could never perform—such as an amulet meant to ward off ghosts. As noted in note 65, phantom functions resemble genuine functions in the epistemic division of labor that they produce. I may receive an amulet from a designer who purports to have a mechanistic explanation for how it performs its supposed function; I can use the amulet in my practical reasoning without understanding how it supposedly performs its function, simply by following use plans communicated to me by the maker. However, the stability or reproduction of the amulet cannot be attributed to its actually warding off ghosts. The present theory suggests a response similar to Preston's (1998) response. Because the persistence theory is a property-oriented, causal-role account, it must explain the stability of these kinds in virtue of things that they *actually* do. For this reason, phantom functions cannot be genuine functions. Nevertheless, artifacts that have phantom functions, such as *fengshui* mirrors and magical amulets, are often stable parts of cultures. In this way, phantom functions suggest the existence of genuine functions, even though they can never be genuine functions themselves: these artifacts may have the function to provide comfort, say, or increased aesthetic value to a house.

This move helps to assimilate the persistence-based view of functions with the social sciences. As noted above, Merton distinguishes between the *manifest* functions and *latent* functions of social entities (see e.g. Merton 1968). While many social practices and institutions are explicitly designed and have transparent meanings for individuals, they also make contributions to the societies in which they exist, which do not always align with what individuals expect. Similarly, we suggest that philosophers should be skeptical of the relationship between human instrumental reason and the functions of artifacts.

All in all therefore: applying presentist social functionalism to artifacts suggests that the function of an artifact type consists those of the features of tokens of this type that increase the expected reproductive or survival success of that type. This is a useful account of artifact functions, as it avoids the pitfalls of the alternative accounts while preserving their benefits—and having some novel positive features on top of that. In particular, the account is non-arbitrary, non-historical, non-intentionalist, and actualist. It also can account for the multiple realizability of artifact functions, the multiple usability of artifacts, their ability to be "recycled," the fact that they can reproduce, malfunction, and the existence of phantom functions. In this way, the account allows us to determine which artifacts are stable parts of society and which merely passing fads: by linking artifacts to patterns of use and social institutions, the study of artifact functions is placed directly at the center of the social science. To make this clearer, the next section briefly considers two concrete cases: off-label drug uses and large language models like ChatGPT.

6.4 Two Case Studies

There are many possible case studies that could be considered to illustrate this theory of artifact functions, but the two chosen here have the benefit of bringing out several further points of interest for the discussion of presentist social functionalism that is central to this book. In particular, they emphasize the interventionist role that the present account can play, as well as making clearer what sort of data we should consider in our social scientific investigations of novel artifacts. There is no question that what follows are not detailed treatments of these very complex issues, but merely brief outlines to bring out the fruitfulness of the theory of artifact functions defended in this chapter.

6.4.1 Pharmaceuticals and off-Label Usage

The pharmaceutical industry is heavily regulated, in the sense that drugs, treatments, or medical devices get approved for a specific set of uses. While medical professionals are free to prescribe them for other uses, such off-label prescriptions may or may not be covered by health insurance, and pharmaceutical companies in the US are not allowed to promote them (Wittich et al. 2012).[9]

For present purposes, there are two key things to note about this practice. First, whatever may be true about the regulatory framework and the research and testing that went into drug approval for purposes P, if a drug becomes sufficiently widely used for a different purpose P′, from the point of view of social science, P′ should be seen to be its function. For a concrete example, consider Ozempic, which was developed and approved to lower blood sugar in patients with type 2 diabetes. However, it has come to be widely used as tool for weight loss (Ozempic for weight loss: Does it work, and what do experts recommend? 2023). If this is true, the social function of this drug should be seen to aid in weight loss. That is, the current theory of artifact function makes clear that, apart from regulatory ways to understand drugs, treatments, and medical devices, there is also a social scientific one that might differ from the former. This is again similar to what was the case in earlier chapters: for example, the legal purposes we have when it comes to classifying something as a "corporation" may differ from our social scientific purposes. The same is true: there may be good medical or public policy-related reasons for seeing the function of a treatment or medical device as whatever it got regulatory approval for; however, there may also be good social scientific reasons to diverge from this, and see this function as lying elsewhere.

Second and relatedly, accepting that Ozempic has the social function of a weight-loss tool, it becomes (pro tanto) reasonable for us to take steps ensuring that this use of the drug is respected and enabled. This may include changing the regulatory frameworks so that this use can be advertised, covered, and perhaps even ensured to be safe and effective; however, it may also include other interventions. Among these is the fact that, if the social function of Ozempic (say) is a tool for weight loss—which potentially has a much higher user-based than patients with type 2-diabetes—steps should be taken that this demand can be met. This may include allowing pharmaceutical companies to produce more doses of the drug than would be expected given the population of type 2-diabtes patients: while this may be inconsistent with the approved label, it is in line with the social function of this artifact. This would do justice to social reality, and also allow type 2-diabetes patients continued access to the medicine.[10]

Overall, the key point to note here is that a presentist social functionalist perspective on artifact functions makes clear that off-label uses of drugs, treatments, and medical devices can become the (social) *function* of these drugs, treatments, and medical devices. In turn, this may require that we shift our interventions in line with this fact. While this is sometimes implicitly recognized—sufficiently widespread

---

[9] If sufficiently widespread, off-label drug uses get compiled in a "compendium," though (Wittich et al. 2012).
[10] Additionally, it *might* mean that the manufacturer has an obligation to ensure that the drug is safe and effective when used for weight loss. However, this introduces many legal, political, and moral complexities that require a separate treatment of their own.

off-label uses can lead to a re-classification of a drug—the treatment defended in this chapter puts this reasoning on a systematic basis.

6.4.2 ChatGPT and Other Large Language Models

ChatGPT and other large language models used for predicting the next set of words in a sequence are currently extremely widely discussed, both in professional and private circumstances. This is not surprising. Technology like this has the potential to disrupt many industries, and indeed many aspects of life as we know it. It becomes possible to generate grammatically correct, semantically sensible (at least generally) text for a variety of purposes in extremely short amounts of time.

However, these models also raise the question of what, exactly, they are *for*? Are these models chatbots? Research tools for linguists or computer scientists? Aids for internet searches? Answering these questions matters, as doing so has implications for how we should treat this new technology. Should it be banned in schools and universities? Should it be encouraged there? Should it be banned or encouraged among government workers? What about among scientists? Are certain uses of ChatGPT and other LLM's—as a therapist, say, or medical provider—to be discouraged, and others—say, as an internet search engine—to be encouraged? Without knowing the function of LLM's like ChatGPT, it is difficult to answer these questions. The presentist social functional perspective can help ground answers to these questions.

To see this, begin by noting that this case clearly illustrates the difficulties of the intentionalist account. While nominally designed as a "chatbot," it is not clear that the designers behind ChatGPT and other LLM's had or have a specific use of the app in mind. Rather, they were driven to build large language models for their own sake and because they may be useful for a variety of reasons. This generality makes it difficult to assess which uses are properly functional and which are not. For example, in 2023, two lawyers in New York were fined for using ChatGPT to generate legal briefs, which contained references to non-existent cases (Weiser 2023, June 22). Is this a failure of the algorithm—should it have provided them with accurate case law—or a misuse of the system—should they not have used a chatbot for creating legal briefs? Given the indeterminateness of the intentions behind the "design" of the algorithm, it becomes difficult to answer this question—*if* intentions are seen to ground the function of the artifact. This again suggests that designers' intentions are simply a poor foundation for ascribing a function to ChatGPT.

Second, recall that the intentionalist account builds on the commonsense association of *functions* with *design*. However, machine learning models like LLM's are largely not designed by humans but produced through an automated optimization process. ChatGPT is based on GPT-3.5 and GPT-4, which were produced using a deep neural network. Partly as a result, such complex models exhibit a great degree of opacity: often, even the "designers" don't know (and perhaps *can't* know) how these algorithms work (Creel 2020). While the designers understand the training algorithms, they often don't understand how the resulting model produces the outcomes that it does. Thus, the designers of such models lack the epistemic authority that designers are typically assumed to have. To the extent that this is true, functional ascriptions to such models thus fail to meet Houkes and Vermaas (2010) support requirement—and we might thus need to conclude that LLM's like ChatGPT have no function.

By contrast, the present account is in a much better position here, and can make clearer the kind of information we need to consider to determine the function of ChatGPT and other LLM's. This information concerns the patterns of use of these LLM's. The function of an LLM consists in those of its effects that contribute to its persistent use, given the selection pressures it currently faces. So, if it turns out that people mostly use ChatGPT and other LLM's to avoid figuring out what to write in response to certain questions—whether these be exams or in work-contexts—then this is its function. Like other tools for cheating, there would then be some justification for banning it in schools—despite the fact that it may have other, more positive uses as well. By contrast, if it were to turn out that people mostly use ChatGPT and other LLM's as a way to quickly obtain information about a variety of issues that they then use in other own, original writing, this use should be encouraged—and this is so even though it can also be used for cheating purposes.

In this way, the presentist social functionalism perspective of artifact functions makes clear what we need to study, from a social scientific point of view, to understand ChatGPT and other LLM's better. While it is too early to tell what the results are of this study, it is at least clear that what we need to do here. In this way, the presentist social functionalism account of artifact functions can be shown to be a fruitful tool for the further investigation of issues of contemporary social importance.

6.5 Conclusion

This chapter used the perspective of presentist social functionalism to develop a new account of artifact functions. According to this account, the function of an artifact types consists in those usage pattern of tokens of the artifact which increase the expected reproductive or survival success of the type in the current social environment. This account improves on the mainstream intentionalist accounts of artifact functions, as it avoids appeal to the sometimes unclear mental states of designers and does justice to the social nature of artifact use. It also improves on the historical reproduction-based account, as its core mechanism is spelled out more clearly. This account can be shown to satisfy the key features any compelling account of artifact functions should satisfy: multiple realizability, multiple utilizability, recycling, reproduction with variation, malfunction, and phantom functions. Finally, the account is shown to have fruitful practical implications: for example, it can suggest novel interventions in the context of off-label drug uses and make clearer which data should be collected to establish the function of ChatGPT and other large language models.

References

Bigelow, J. C. 1998. Functionalism in Social Science. In *Routledge Encyclopedia of Philosophy*. Taylor and Francis.

Bijker, W., and T. J. Pinch. 1987. The Social Construction of Facts and Artifacts: Or How the Sociology of Science and the Sociology of Technology Might Benefit Each Other. In *The Social Construction of Technological Systems*, ed. W. Bijker, T. Hughes, and T. Pinch. MIT Press.

Boyd, R., and P. Richerson. 1985. *Culture and the Evolutionary Process*. Chicago: University of Chicago Press.

Creel, K. A. 2020. Transparency in Complex Computational Systems. *Philosophy of Science* 87 (4): 568–589.

Davis, J. L. 2020. *How Artifacts Afford: The Power and Politics of Everyday Things*. Cambridge, MA: MIT Press.

Dipert, R. 1993. *Artifacts, Art Works, and Agency*. Philadelphia: Temple University Press.

Dipert, R. 1995. Some Issues in the Theory of Artifacts: Defining 'Artifact' and Related Notions. *The Monist* 78 (2): 119–135.

Evnine, S. J. 2016. *Making Objects and Events: A Hylomorphic Theory of Artifacts, Actions, and Organisms*. Oxford: Oxford University Press.

Feenberg, A. 1999. *Questioning Technology*. London: Routledge.

Franssen, M. 2006. The Normativity of Artefacts. *Studies in History and Philosophy of Science A* 37 (1): 42–57.

Franssen, M., P. Kroes, T. Reydon, and P. E. Vermaas. 2014. *Artefact Kinds: Ontology and the Human-Made World*. Springer.

Garson, J. 2019. *What Biological Functions are and Why They Matter*. Cambridge: Cambridge University Press.

Hilpinen, R. 1993. Authors and Artifacts. *Proceedings of the Aristotelian Society* 93:155–178.

Holm, S. 2017. The Problem of Phantom Functions. *Erkenntnis* 82 (1): 233–241.

Houkes, W. 2006. Knowledge of Artefact Functions. *Studies in History and Philosophy of Science A* 37 (1): 102–113.

Houkes, W., and P. Vermaas. 2004. Actions Versus Functions: A Plea for an Alternative Metaphysics of Artifacts. *The Monist* 87 (1): 52–71.

Houkes, W., and P. Vermaas. 2010. *Technical Functions: On the Use and Design of Artefacts*. Dordrecht: Springer.

Lachney, M., and T. Dotson. 2018. Epistemological Luddism: Reinvigorating a Concept for Action in 21st Century Sociotechnical Struggles. *Social Epistemology* 32 (4): 228–240.

Latour, B. 1990. Technology is Society Made Durable. *The Sociological Review* 38 (1): 103–131.

McLaughlin, P. 2000. *What Functions Explain: Functional Explanation and Self-Reproducing Systems*. Cambridge: Cambridge University Press.

Merton, R. 1968. *Social Theory and Social Structure*. New York: Free Press.

Millikan, R. 1984. *Language, Thought, and Other Biological Categories*. The MIT Press.

Mumford, L. 1964. Authoritarian and Democratic Technics. *Technology and Culture* 5 (1): 1–8.

Naoe, K. 2008. Design Culture and Acceptable Risk. In *Philosophy and Design: From Engineering to Architecture*, ed. P. E. Vermaas, P. Kroes, A. Light, and S. Moore, 119–130. Springer.

Norman, D. 2013. *The Design of Everyday Things: Revised and Expanded Edition*. New York: Basic Books.

Ozempic for weight loss: Does it work, and what do experts recommend? (2023, July 19). UC Davis Health. https://health.ucdavis.edu/blog/cultivating-health/ozempic-for-weight-loss-does-it-work-and-what-do-experts-recommend/2023/07#:~:text=Ozempic%20works%20by%20mimicking%20a,the%20effect%20of%20bariatric%20surgery

Polger, T., and L. Shapiro. 2016. *The Multiple Realization Book*. Oxford: Oxford University Press.

Preston, B. 1998. Why Is a Wing Like a Spoon? A Pluralist Theory of Function. *The Journal of Philosophy* 95 (5): 215–254.

Preston, B. 2009. Philosophical Theories of Artifact Function. In *Philosophy of Technology and Engineering Sciences*, ed. A. Meijers, 213–233. North-Holland.

Preston, B. 2013. *A philosophy of Material Culture: Action, Function, and Mind*. London: Routledge.

Scheele, M. 2006. Function and Use of Technical Artefacts: Social Conditions of Function Ascription. *Studies in History and Philosophy of Science A* 37 (1): 23–36.

Shapiro, L. 2004. *The Mind Incarnate*. Cambridge, MA: MIT Press.

Van Eck, D., and E. Weber. 2014. Function Ascription and Explanation: Elaborating an Explanatory Utility Desideratum for Ascriptions of Technical Functions. *Erkenntnis* 79 (6): 1367–1389.

Vermaas, P., and W. Houkes. 2003. Ascribing Functions to Technical Artefacts: A Challenge to Etiological Accounts of Functions. *The British Journal for the Philosophy of Science* 54 (2): 261–289.

Walsh, D., and A. Ariew. 1996. A Taxonomy of Functions. *Canadian Journal of Philosophy* 26 (4): 493–514.

Weiser, B. (2023, June 22). *ChatGPT Lawyers are Ordered to Consider Seeking Forgiveness*. The New York Times. https://www.nytimes.com/2023/06/22/nyregion/lawyers-chatgpt-schwartz-loduca.html

Weiskopf, D. 2011. The Functional Unity of Special Science Kinds. *The British Journal for the Philosophy of Science* 62 (2): 233–258.

Winner, L. 1980. Do Artifacts Have Politics? *Daedalus* 109 (1): 121–136.

Wittich, C. M., C. M. Burkle, and W. L. Lanier. 2012. Ten Common Questions (and their answers) About Off-Label Drug Use. *Mayo Clinic Proceedings* 87 (10): 982–990. https://doi.org/10.1016/j.mayocp.2012.04.017.

Wright, L. 1976. *Teleological Explanations: An Etiological Analysis of Goals and Functions*. Berkely: University of California Press.

Ziman, J., ed. 2000. *Technological Innovation as an Evolutionary Process*. Cambridge University Press.

Open Access This chapter is licensed under the terms of the Creative Commons Attribution 4.0 International License (http://creativecommons.org/licenses/by/4.0/), which permits use, sharing, adaptation, distribution and reproduction in any medium or format, as long as you give appropriate credit to the original author(s) and the source, provide a link to the Creative Commons license and indicate if changes were made.

The images or other third party material in this chapter are included in the chapter's Creative Commons license, unless indicated otherwise in a credit line to the material. If material is not included in the chapter's Creative Commons license and your intended use is not permitted by statutory regulation or exceeds the permitted use, you will need to obtain permission directly from the copyright holder.

Chapter 7
Conclusion

Abstract This chapter summarizes the overall argument of the book, draws out some general themes, and points in the direction of further work on this topic. The chapter also emphasizes the methodological innovations of the book as a whole: in particular, it shows how the book brings together cutting-edge work in biological sciences with that in the social sciences in an application-focused context.

Social institutions matter: they are the rules and norms that help structure human social behavior. Unsurprisingly, therefore, they are a key focal point of research in the social sciences. One of the major aspects of this research concerns the question of which of the myriad social institutions that make up a given society—from marriage conventions to communication protocols and religious rituals—are stable, foundational features of that society, and which merely passing fads that will disappear as quickly as they appeared. A traditional way of addressing this question is by considering the *function* of a social institution. However, (in)famously, determining what the function of a social institution is, and what roles functional attributions do and should play in the social sciences is far from easy and has led to vigorous debate. This book tried to push this debate forward by using recent insights from the biological and cognitive sciences to develop and apply a new account of social functionalism. There are several important upshots of the application-focused discussion here presented that deserve to be made explicit.

First, the best approach towards social functionalism is presentist, actualist, non-arbitrary, and general. According to this approach, the function of a social institution is constituted by those of its features that increase its expected survival or reproductive success in the current socio-cultural environment. This account incorporates recent insights from (philosophy) of biology, such as the possibility to make sense of natural selection even in populations of entities that do not reproduce, and even where that population only contains one such entity.

Second, presentist social functionalism can be shown to satisfy the desiderata left open by the alternative approaches: structural-functionalism, historical-selectionist accounts, virtual-selectionist accounts, and intentional-design-based accounts. It can also stand up to some key objections that can be raised against it, such as the fact that it is too hard to apply or that it answers the wrong kinds of questions. For example, while it is true that the account requires some individuation schema of institutions and features, it can help bootstrap such an individuation schema. More generally, it is important to emphasize that no account of social functionalism should be expected to be able to answer all questions in the social sciences. While—as just noted—the functionalist perspective is indeed a key tool in the social scientific toolbox, it should not be seen as the only such tool.

The third key point to note here concerns the fact that presentist social functionalism is a powerful addition to work in the social functionalist tradition, as can (i) open up questions that have not been addressed before, (ii) provide a foundation for theoretical insights in other areas, and (iii) be extended into neighboring areas. In particular, a presentist social functionalist approach to the question of the function of corporations suggests that different corporations may have different functions, which may furthermore by dynamically changing and be falling in between the traditional stakeholder-shareholder dichotomy. Furthermore, present social functionalism can be used as the backbone on which to hang a novel account of institutional corruption, according to which an institution is corrupted if the operation of its functional features is blocked, and where these functional features are spelled out in terms of presentist social functionalism. Presentist social functionalism also allows for the formulation of a new theory of artifact functions, according to which the function of an artifact types consists in those usage pattern of tokens of the artifact which increase the expected reproductive or survival success of the type in the current social environment. In short: presentist social functionalism is not just philosophically interesting; it has direct implications for ongoing debates in the social sciences.

The final point to make explicit here concerns not presentist social functionalism itself, but rather the methodology of its development. In this book, the development of presentist social functionalism goes hand-in-hand with its application. In this way, the book can also be seen as a case study for an applied form of philosophy of (social) science. To make clear how a novel theoretical perspective works, what its benefits and drawbacks are, how fruitful and clear it is, it can be useful to actually put the approach to work in addressing some existing controversies in the (social) sciences. In this way, it is hoped that the book can also be of value for scholars not directly concerned with social functionalism.

All in all, therefore: this book looks to set out an exciting, novel account of social functionalism, show why the account is compelling, apply it to some existing debates in the social sciences, and thus present a case study of an application-based philosophy of (social) science. In this way, it is hoped that the book makes for a step forward in the long-standing debate surrounding the functions of social institutions.

Open Access This chapter is licensed under the terms of the Creative Commons Attribution 4.0 International License (http://creativecommons.org/licenses/by/4.0/), which permits use, sharing, adaptation, distribution and reproduction in any medium or format, as long as you give appropriate credit to the original author(s) and the source, provide a link to the Creative Commons license and indicate if changes were made.

The images or other third party material in this chapter are included in the chapter's Creative Commons license, unless indicated otherwise in a credit line to the material. If material is not included in the chapter's Creative Commons license and your intended use is not permitted by statutory regulation or exceeds the permitted use, you will need to obtain permission directly from the copyright holder.

The manufacturer's authorised representative in the EU is Springer Nature Customer Service Centre GmbH, Europaplatz 3, 69115 Heidelberg, Germany. If you have any concerns regarding our products, please contact ProductSafety@springernature.com

Printed and bound by CPI Group (UK) Ltd, Croydon, CR0 4YY

26/03/2026

02078942-0011